Crystal Research and Technology

9

Journal of Experimental and Industrial Crystallography

Zeitschrift für experimentelle und technische Kristallographie

Volume 22
1987
Number 9 September

THEORIA CUM PRAXI

Akademie-Verlag · Berlin

ISSN 0232-1300 Cryst. Res. Technol., Berlin 22 (1987) 9, 1115–1222, K141–K160

Directions to the author

In the journal "Crystal Research and Technology – Kristall und Technik" as a rule the following categories of papers will be printed: original papers, review articles, short notes, and book reviews, **preferentially in English**. Complete information necessary for type-setting and a "Short Title" should be added.

The manuscript should not exceed 10 pages and be type-written on one side only in the format A4 (21 × 29.7 cm) on white ink-proof paper, with a two-lined distance and a left-hand free edge of 3 cm, and submitted in 2 copies. All sheets of paper belonging to the contribution are continuously to be numbered.

The head of the manuscript should be arranged in this manner: initials and name of the author; institution, and title of the paper. Then two abstracts should follow, one in the original language and the second in German, English, French, or Russian, respectively.

Formulae, signs and symbols, letters etc. being wanting on the type-writer, should be distinctly written with penholder. O as zero is to be underlined, the capital O is not. Greek letters should be underlined with red ink.

Paragraphs should be marked by indenting of at least 3 touches.

Concerning the marking of the manuscript should be noted:
headings should be underlined, sub-headings not, and indented by 9 touches. The titles of tables are not to be underlined. Other markings in the text will be done by the editorial staff. They mean:

light-red ∿∿∿∿∿∿∿ = small capitals
blue ∿∿∿∿∿∿∿ = in italics

light-red | perpendicular straight red line on the left edge = small printing

Tables are to be added on separate sheets of paper at the end of the contribution.

Foot notes should be continuously written on a separate sheet of paper.

Figures (Fig.) – drawings and photographs – should be not more than 8 and submitted (2 copies) in a special envelope: photographs as white high-gloss prints (inscription on a transparent covering sheet) and drawings as distinct sketches. On the back of the submitted figures the reduction scale in tenth (9/10, 8/10, 5/10 e.g.), author's name, short title of the paper, and the number of the figure should be given.

Captions for the figures should be submitted on a separate sheet of paper.

References should be written without numbers in alphabetical order of the names at the end of the contribution. Cited names in the text stand for the references. The references should contain the following items, for example:

for books:	RAST, K.: Physikalisch-chemische Rechnungen, Berlin 1958
for journals:	KRAUSE, H. N., MÜLLER, K.: Cryst. Res. Technol. **16** (1981), 70

Add, please, the short title, about 6 keywords, and the academic grades of the authors.

The authors will be sent two galley proofs of review articles and original papers. Proof reading of short notes and book reviews will be done by the editorial staff.

For each of the papers a total of 50 reprints is available free of charge. Requests for additional maximum 200 reprints (against payment) should be given together with the submitted manuscript.

9

Crystal Research and Technology

Founded 1966 as
Kristall und Technik

Journal of Experimental and Industrial Crystallography

VOLUME 22 · 1987 · NUMBER 9

Akademie-Verlag · Berlin

Papers should be directed to the editor's office
DDR-7030 Leipzig, Scharnhorststraße 20,
or to the editor Prof. Dr. Hermann Neels,
DDR-7254 Machern, Zeititzer Weg 25,
or to one of the members of the Advisory Board.

Orders can be sent

- in the GDR:
  to the Postzeitungsvertrieb or to the Akademie-Verlag Berlin, Leipziger Straße 3–4, DDR-1086 Berlin;
- in the other socialist countries: to a book-shop for foreign language literature or to the competent news-distributing agency;
- in the FRG and Berlin (West): to a book-shop or to the wholesale distributing agency Kunst und Wissen, Erich Bieber OHG, Wilhelmstr. 4–6, D-7000 Stuttgart 1;
- in the other Western European countries: to Kunst und Wissen, Erich Bieber GmbH, General Wille-Str. 4, CH-8002 Zürich;
- in USA and Canada to the VCH Publishers, Inc., 303 NW 12th Avenue, Deerfield Beach, FL 33442-1788, USA;
- in other countries. to the international book- and journal-selling trade, to Buchexport, Volkseigener Außenhandelsbetrieb der DDR, Postfach 160, DDR-7010 Leipzig, or to the Akademie-Verlag Berlin, Leipziger Straße 3–4, DDR-1086 Berlin.

Journal „Crystal Research and Technology"

Herausgeber: Im Auftrage des Akademie-Verlages Berlin herausgegeben von Prof. Dr. Hermann Neels.

Verlag: Akademie-Verlag Berlin, Leipziger Str. 3–4, DDR-1086 Berlin; Fernruf: 2236221 oder 2236229; Telex-Nr.: 114420; Bank: Staatsbank der DDR, Berlin, Kto.-Nr. 6836-26-20712.

Chefredakteur: Prof. Dr. Hermann Neels, Zeititzer Weg 25, DDR-7254 Machern; Fernruf: Brandis 3442.
Anschrift der Redaktion: Scharnhorststr. 20, DDR-7030 Leipzig; Telefon: 310502.

Veröffentlicht unter der Lizenznummer 1526 des Presseamtes beim Vorsitzenden des Ministerrates der Deutschen Demokratischen Republik.

Gesamtherstellung: VEB Druckerei „Thomas Müntzer", DDR-5820 Bad Langensalza.

Erscheinungsweise: Das Journal „Crystal Research and Technology" erscheint monatlich. Die 12 Hefte eines Jahrganges bilden einen Band. Bezugspreis je Band 480,– DM zuzüglich Versandspesen; Preis je Heft 40,– DM. Der gültige Jahresbezugspreis für die DDR ist der Postzeitungsliste zu entnehmen.

Bestellnummer dieses Heftes: 1082/22/9.

AN (EDV) 35830
02800

Cryst. Res. Technol.	22	1987	9	1117–1119

Original Papers

C. C. DESAI

Department of Physics, Sardar Patel University, Vallabh Vidyanagar 388120, Gujarat State, India

Theoretical Aspects of the Crystal Growth in Gel Medium

Crystal growth by diffusion process in a gel medium has been utilized for a long time as method of simulation processes in the laboratory. The suitable theory of crystal growth in gels has been developed. Attention has been made mainly to the more general case in which the counter-diffusion of two reacting substances occurs. The implications are discussed.

1. Introduction

The gel technique to day, resting on a sound footing, provides an important elegant tool for studying kinetics and mechanism of bulk crystallization as affected by various parameters such as gel *p*H, gel aging, gel quality, gel density, environmental temperature, nature and concentration of acid, purity and strength of reactants (supernatants) type of crystallization apparatus, etc. The gel *p*H has been found to behave as habit modifiers (DESAI, RAI). The concentration of reactants (DESAI, HANCHINAL) and temperature (CHARMEL) frequently modify the crystal size and morphology of the grown crystals. The aim of this paper is to give recent advances in theory of crystal growth in gels.

2. Theoretical aspects of growth mechanism

As regards the mechanism by which crystals grow in the silica gel medium, there have been evidences to two-dimensional piling and spreading of growth layers predominantly taking place (DESAI, RAI) from one or more initiation centres. One is, therefore, tempted to make analogical comparison with solution growth. When single crystals of slightly soluble salts are grown from highly supersaturated solutions, which just remain stable over long periods by slow evaporation, or cooling, these crystals never become larger ones, because their supply of dissolved compound to the crystal surfaces is still extremely feeble. The fact that the large crystals are generated in gels which proves that the supersaturation near the surface of these crystals must be much higher in gel than in solution. This suppression prevents the development of heterogeneous nuclei. So only homogeneous nucleation is left which occurs at high supersaturation only. What in fact goes on in the medium is the diffusion of dissolved materials as a consequence of the causal character of the thermal motion of the molecules.

The molecules pass through an energy barrier, ΔG, the free energy of activation for the molecular transport process. A molecule or an ion changes its place with a frequency

$$f = T^{-1} = \frac{kT}{h} \exp\left(-\frac{\Delta G}{kT}\right), \tag{2.1}$$

where

k Boltzmann constant
T growth temperature
h Planck's constant

But according to EINSTEIN the root mean square displacement (λ) in time T is

$$\lambda = (2DT)^{1/2} \tag{2.2}$$

where D is the diffusion constant.

Putting $\lambda = d$, the ionic diameter, the frequency of nucleation is

$$f = T^{-1} = \frac{2D}{d^2} = \frac{kT}{h} \exp\left[-\frac{\Delta G}{kT}\right]. \tag{2.3}$$

The energy (ΔG) is equivalent to E_c, the energy available for the creation of crystal nucleus, given by

$$E_c = \frac{16}{3} \frac{\sigma^3 \lambda}{\varrho^2 L^2}, \tag{2.4}$$

where

σ surface energy per unit area
ϱ droplet density
L heat of dissolvation

According to Fick's law:

$$J = -D \operatorname{grad} C \tag{2.5}$$

and

$$\frac{dC}{dt} = D \nabla^2 C, \tag{2.6}$$

where J is the quantity of matter moving in the direction of the concentration gradient per unit area perpendicular to ∇C. Assuming that the precipitates are not too close to one another, as is true of gel growth, the concentration in space around the crystal of radius R is a function of the distance x from the centre and the boundary conditions become

$$\begin{aligned} C(x, 0) &= C_\infty \\ C(R, t) &= S \\ C(\infty, t) &= C_\infty . \end{aligned} \tag{2.7}$$

The solution of diffusion equations (2.5) and (2.6) gives the radius R as a function of the speed V of the advancing growth front in the gel medium:

$$R = [DV(C_\infty - S)\, t]^{1/2} . \tag{2.8}$$

Since each particle is to be treated independently of the others, the factor $V(C_\infty - S)$ is small and hence equation (2.8) reduces to

$$R^2 = (Dt)^{1/2} \tag{2.9}$$

where

D diffusion constant and
t time of nucleation.

The equation (2.9) which is typical of one-dimensional diffusion process has been experimentally verified by Blank, Liaw and Faust, and Desai and Rai. This equation suggests that the chemical reaction is not the rate-determining process in crystal growth and the existence of a constant supersaturation in the crystal surface.

2.1. Growth rate constant

Relation between the length of the crystals and the gel aging time or the growth temperature can be expressed by the specific mathematical functions

$$r^2 = kt \tag{2.10}$$

and

$$r^2 = k + Dt\,, \tag{2.11}$$

where

r length of the crystal
t growth period/growth temperature
k growth rate constant ($mm^2 \cdot h^{-1}$)

It has been shown that the parabolic kinetics are characteristics of a one-dimensional diffusion-controlled process. The equation (2.11) suggests that the chemical reaction was not the rate determining process in crystal growth and the existence of a constant supersaturation in the crystal surface. However, it is important to note that some deviations from relation (2.11) have been found for extreme t values of suggesting different growth mechanism of earlier and later stages of the crystal growth.

The author is thankful to University Grants Commission (New Delhi) for the sanction of the project, entitled, "Studies of growth, characterization, and properties of alkali halides and tartrate single crystals".
CCD is very much grateful to Mr. Arvind Desai, Mrs. Lora, Miss Gladys and Lorraine (London, U.K.) for their interest and encouragement.

References

Blank, Z.: J. Cryst. Growth **18**, 281 (1973)
Charmel, J. F.: J. Cryst. Growth **8**, 291 (1971)
Desai, C. C., Rai, J. L.: J. Cryst. Growth **50**, 562 (1980)
Desai, C. C., Hanchinal, A. N.: Cryst. Res. Technol. **18**, 1499 (1983)
Desai, C. C., Rai, J. L.: Cryst. Res. Technol. **16**, 701 (1981)
Einstein, A.: Ann. Physik **19**, 371 (1906)
Liaw, H. M., Faust, J. W., Jr.: J. Cryst. Growth **8**, 8 (1971)

(Received February 1, 1987)

Author's address:

Professor C. C. Desai
Department of Physics, Sardar Patel University
Vallabh Vidyanagar 388120 Gujarat State, India

Book Review

E. I. GIVARGIZOV

Highly Anisotropic Crystals

Series MSMR (Material Science of Minerals and Rocks)

D. Reidel Publishing Company, Dordrecht, Boston, Lancaster, Tokyo in co-publication with Terra Scientific Publishing Company, Tokyo, 1987

394 Seiten mit 100 Abbildungen und 21 Tabellen, Sachwörterverzeichnis, Substanzverzeichnis, Preis 230,– Dfl., 98,– US Dollar. ISBN 90-277-2172-6

In den 50-er Jahren erschienen erste gezielte Arbeiten zur Interpretation einer besonders ausgeprägten Wachstumsform von Kristallen, die zumeist als Begleiterscheinung bei Sublimations- und Aufdampfprozessen beobachtet und morphologisch mit sehr dünnen Nadeln (Whiskers) verglichen werden. Seither existieren weit über 1000 Originalarbeiten zu Untersuchungen eines solchen Habitus. Fast alle kristallinen Stoffe neigen unter bestimmten Bedingungen zu diesen und ähnlichen Fassons und es fehlt weltweit nicht an Versuchen, diese beherrschbar für gezielte spezielle Anwendungen zu gestalten. E. I. GIVARGIZOV widmet sich dieser Problematik sowie der whiskersgestützten Plättchenzüchtung seit über 15 Jahren und gilt als einer der führenden Spezialisten auf diesem Gebiet. Die vorliegende Monographie greift einerseits auf bekannte Betrachtungen der ersten Zusammenfassung zurück und erweitert andererseits um zahlreiche weitere Fragestellungen, wie das Wachstum von Whiskers, Plättchen, Hohlkörpern, Dendriten, interkalarer Strukturen u. a. nicht nur aus dem Dampf, sondern auch aus der Schmelze, Lösungen und festen Phasen. Gleichzeitig werden Befunde zu ausgeprägt anisotropen natürlichen Wachstumsformen von Mineralen, wie das Nadel- oder Plättchenwachstum von Silikaten mit unendlicher Ketten- oder Schichtenstruktur u. a. diskutiert. Eine weitere interessante Bereicherung erfährt das vorliegende Buch durch die Einbeziehung anisotroper Gefügestrukturen (Eutektika) und Oberflächentexturierungen. Somit wird eine Verallgemeinerung bei der Betrachtung einer wesentlichen Eigenart der Kristalle — dem anisotropen Wachstum — gegenüber amorpher Stoffe erzielt.

Die Monographie ist in 8 Teile mit eigenständigen Literaturanhängen untergliedert. Übergreifenden Charakter besitzt die Diskussion der entscheidenden drei Faktoren für ein hochanisotropes Kristallwachstum, sei es in Form von Whiskers, Fäden, Plättchen, Dendriten oder Eutektika und Texturen — die kristallchemische Struktur der Kristalle und ihre Symmetrie, das Wachstumsmilieu und die Art dessen Einwirkung und die Wachstumskinetik mit ihrer Differenziertheit über die einzelnen kristallographischen Flächen. Einen zentralen Punkt nehmen spureninduzierte VLS-Whiskers ein. Es wird deutlich, daß Whiskers und andere hochanisotrope Wachstumsformen und Strukturen ein besonders geeignetes Objekt zur Klärung prinzipieller Fragen des Kristallwachstums darstellen, da sie zumeist hochperfekt kristallisieren. Das gilt für die Einbaukinetik, das Löslichkeitsverhalten und die Segregation von Verunreinigungen, die Diffusion, das Stabilitätsverhalten u. a. Es lassen sich — in einem speziellen Teil der Darstellung zusammengefaßt — interessante Anwendungsmöglichkeiten ableiten: Feldemissionskathoden, Dehnmeßgeber, Temperaturmeßgeber etc. Von besonderem Interesse dürften geordnete Whiskersanordnungen für die Mikro- und Optoelektronik oder aber in irregulärer Anordnung auf Werkzeugoberflächen für die Werkstoffbearbeitung sein. Weitere Möglichkeiten eröffnen sich mit den anisotropen Oberflächentexturen zellularer, bacularer oder facettierter Struktur, erzeugt durch Ionenätzen, Sputtern, Festkörperreaktionen und andere Verfahren, auch in Kombination mit der Multiwhiskererzeugung.

Auf Grund seines Überblickcharakters und außerordentlich interessanten Darstellungsweise — die Abbildungen tragen bisweilen spektakulären Charakter — ist das vorliegende Buch für einen breiten Kreis von Spezialisten der Kristallographie, Mineralogie, Metallurgie, Chemie, Physik und Elektronik empfehlenswert.

P. RUDOLPH

Cryst. Res. Technol.	22	1987	9	1121–1126

L. Waliś, H. Jaskólska, Cz. Janusz

Institute of Nuclear Chemistry and Engineering, Warsaw, Poland, and
Institute of Electronic Materials Technology, Warsaw, Poland

Investigation of the Single Crystal Growing Process of Neodymium-doped YAG Garnets Using Radiotracers

Radioisotopic methods of investigation of Nd behaviour during pulling of single crystals of YAG-type garnets were applied. For the investigation of homogeneity of Nd distribution an autoradiographic method was used, and for the determination of Nd contents in single crystals and Nd effective partition ratios — techniques of radiotracers and gamma spectrometry. The conclusions from this work can find technological application.

Introduction

A proper choice of parameters for the process of oxide materials monocrystallization is possible when information is available (among others) concerning the influence of these parameters on the homogeneity of admixture distribution in single crystals (Bradsley, Cockayne; Kestigian, Holloway; Cockayne; Brandle).

Results of many experiments carried out for several years suggested that radiotracer and autoradiographic methods could be very useful for the investigations. For this reason both methods were applied for the control of the cours of the single crystal growing process in the case of $Y_3Al_5O_{12}$ garnets doped with Nd_2O_3, which were prepared for the laser resonator production. The single crystal growing process was performed by the Czochralski method. For the charge heating a high frequency generator (350 to 430 kHz) was used. Its power output accuracy was $\pm 1\%$ and the rotation control accuracy was $\pm 0.15\%$.

Single crystals were pulled from iridium crucibles. As a tracer we used the radioisotope ^{147}Nd. Its nuclear characteristics is given in Table 1. The investigations described below included elaboration of a suitable experimental procedure.

Table 1

Nuclear data for ^{147}Nd

Abundance of matrix isotope	cross section for activ. with thermal neutr.	half-life $T_{1/2}$	E_γ (main)	E_β (main)
17.1%	1.4 barns	11.1 d	91 keV	825 keV

Autoradiographic examination of neodymium distribution in a single crystal

One requirement of a successful autoradiography is that the blackening of the autoradiogram is solely connected with the presence of the studied element.

Two techniques of neodymium labelling were investigated. The first consisted in adding the radioactive isotope ^{147}Nd (as oxide) to the initial charge material for monocrystallization. In the second one, the single crystal pulling was carried out without labelling the charge material and then fragments of the single crystals were irradiated in the nuclear reactor. Both methods can give erroneous results, unless definite conditions are fulfilled.

In the method of ^{147}Nd addition autoradioagrams should be made after a suitable cooling period, because during irradiation at the thermal neutron flux some short-lived radioisotopes are created. Their half-lives are shorter than that of ^{147}Nd, which is 11.1 days, and they blacken the autoradiographuc film similarly as ^{147}Nd. However, the cooling time cannot be too long, because ^{147}Nd decays to ^{147}Pm, which also is a radioisotope and blackens the autoradiographic film. This isotope is a beta emitter similarly as ^{147}Nd, but its half-life is significantly longer, equal to 2.6 years. After a relatively long cooling period of the sample ^{147}Pm radioactivity becomes higher than that of ^{147}Nd. In such case the autoradiogram blackening would indicate the distribution of prometium rather than that of neodymium. Taking into account the above-mentioned factors the following procedure was established as most suitable:

1. after irradiation in the nuclear reactor neodymium oxide (tracer) was cooled during several days;
2. the tracer was added to the charge material and the monocrystallization process was carried out;
3. after cooling (up to 30 days), the autoradiograms were made.

In the second technique of labelling involving irradiation of some parts of the single crystal in the nuclear reactor it was also necessary to use a suitable cooling period, because the concentration of yttrium, being one of the basic components of YAG-garnets, is by two orders of magnitude higher than that of neodymium, and its radioactivity is higher by a few orders of magnitude.

Fortunately, the half-life of ^{90}Y is relatively short, so it is possible to cool it to such level, that in practice it becomes negligible.

Radiometric determination of neodymium concentration and distribution

Preparation of ^{147}Nd radiotracer of a suitable specific activity is an important problem from the point of view of the radiometric measurements. It has been experimentally stated, that the suitable specific activity of neodymium should be in the range from 0.12 to 0.24 MBq/g (about 3.3—6.6 μCi/g) and the tracer ($Nd_2^*O_3$) content of furnace charge material should be about 0.1%.

Procedure

Neodymium oxide in gram amounts was irradiated at a thermal neutron flux of 10^{12} n cm^{-2} s^{-1} in the nuclear reactor. Three 5 mg samples were taken for use as standards and the tracer was added to the furnace charge material. Then the single crystal growing process was carried out. The main parameters, which were sometimes changed in the course of the process were in principle the following: the growth rate of the crystal 0.7 mm/h—1.5 mm/h, the rotation rate $6\ r \cdot min^{-1}$—$80\ r \cdot min^{-1}$. Two devices for the single crystal growing process were used. The difference between them consisted mainly in various thermal equipment.

Samples of the grown crystal were taken in the form of wafers cut out from its various parts. The wafers were carefully polished, cleaned and used for making autoradiograms on an AF-3 film. Then the wafers were crushed and finely powdered (when powdering is not thorough, a significant error arrises during radiometric measurements). The weight of a single sample was about 50 mg. Also from the material remaining in the crucible several samples were taken and powdered. Radioactivity of all samples and standards was measured by means of the Ge(Li) detector of the active volume equal to $26\ cm^3$, and gamma spectrometer IN-90. The counting frequency of the 91 keV gamma radiation was registered.

Results

Nearly all autoradiograms indicate the zone segregation of neodymium (Fig. 1). This segregation reveals the shape of the crystallization front. The frequency of occurrence of zones with higher neodymium concentration is variable. The higher the frequency, the lower the concentration differences within the zones.

The shape of the crystallization front is strongly affected by the process parameters, their stability and the type of the thermal arrangement. At the beginning of the process the parameters are stile not set, so the neodymium distribution of the beginning part of the single crystal is more heterogeneous than that of the rest of the crystal (Fig. 2).

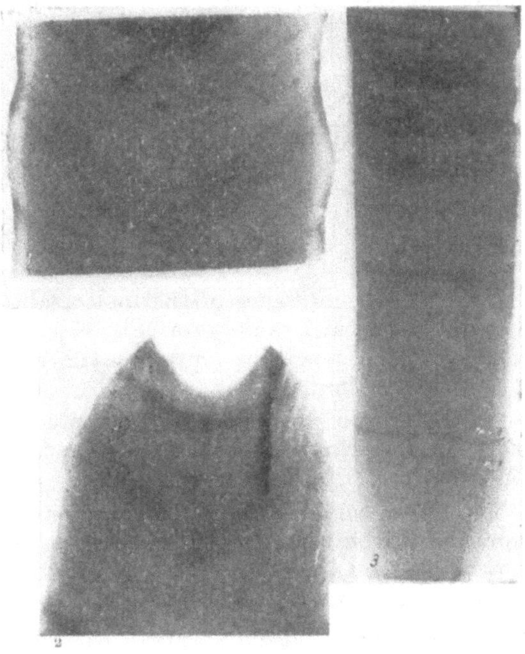

Fig. 1. Autoradiogram of the plate cut out along its vertical axis (the middle part)

Fig. 2. Autoradiogram of the plate cut out along its vertical axis (the top part)

Fig. 3. Autoradiogram of the plate, cut out along its vertical axis (the whole plate)

When the process parameters are not properly chosen, there is a marked heterogeneity of the neodymium distribution.

In the opposite case (suitable parameters) the distribution is more homogeneous, the shape of the crystallization front becomes more flat (the quality of the single crystal is better in that case (BRANDLE)). Moreover, in one case we have obtained an inverse shape (Fig. 3).

In addition to the mentioned zone segregation, local changes in the neodymium concentration were revealed. They appear on autoradiograms as radically distributed blackeninges, perpendicular to the dark bands (Fig. 4, 5). Presumably, these local changes were created due to the existence of imperfections in the single crystal. A wafer cut out from that part of the single crystal, where the imperfections were localized earlier by other methods, showed on the autoradiogram the highest number of imperfections of this type (Fig. 5).

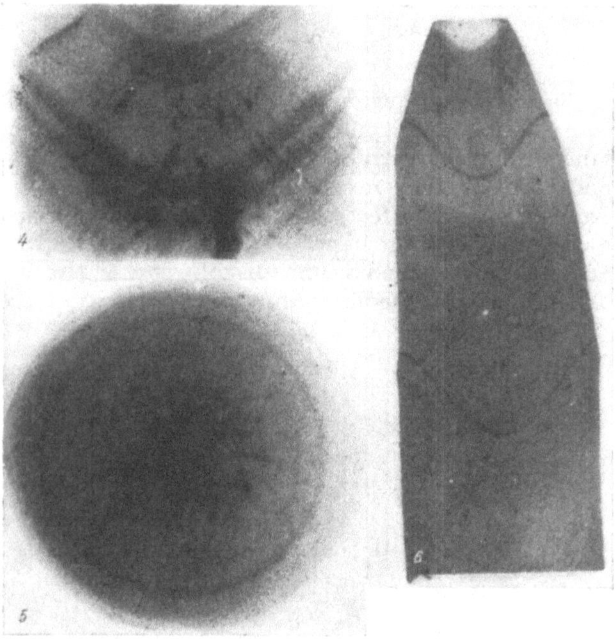

Fig. 4. Autoradiogram of the plate, cut out along its vertical axis (the middle part)

Fig. 5. Autoradiogram of the plate cut out perpendicularly to its vertica axis

Fig. 6. Autoradiogram of the plate, cut out along its vertical axis (the whole plate)

A very great influence of thermal conditions on the neodymium distribution was stated. On the autoradiogram (Fig. 6) the influence of two times introduced small changes of the generator power on the neodymium distribution is shown it a form of two darker lines.

On the basis of numerous experiments we have stated, that for our equipment the most suitable parameters are: the growth rate — 1 mm/h and the rotation rate — 30—50 $r \cdot min^{-1}$.

The tracer contents of the single crystal wafers and of the residue materials were determined from the relative specific activites of the samples and standards.

In order to check the results, balances of relative activities were made in two ways (based on the specific activities of standards, in the second case — on the specific activities of the samples) and their results compared.

The radiometric and autoradiographic results agreed. For instance they showed an increase of the neodymium concentration along the vertical axis from the beginning to the end of the crystal.

From neodymium concentrations in the single crystal, in the residue material and in that at the beginning, one can calculate an approximate value of the effective partition ratio of neodymium (K_{eff}) between the solid phase (single crystal) and the liquid phase (material in crucible). As the neodymium concentration in the liquid phase, the mean value was taken ($K_{eff\ mean}$), calculated from the neodymium concentration in the material in the crucible before and after completing of the processs. As the neodymium concentration in the solid phase we took the mean value of the concentration determined for a whole crystal. In all calculations the concentrations were expressed in the weight percents. The $K_{eff\ beg}$ values were also calculated, and as the neodymium concentration of the liquid phase the concentration of that element in the charge material at the beginning of the process was taken.

Several methods of the partition ratio calculation are described in the literature, for instance by MONCHAMP and BETT and SMIRNOVA et al. They are more accurate, but also more time-consuming. We calculated only the approximate values, however, they were both useful and sufficient for our aims. They enabled to control the neodymium content in the single crystal by suitable choosing the content of this element in the initial material. Also, in the case of pulling a few single crystals, one after another, from one basic charge material, the knowledge of K_{eff} makes possible the preparation of an oxide mixture with the proper neodymium concentration for complementing the charge material after the end of each pulling. Then, the neodymium content of all single crystals pulled in one cycle can be nearly the same. This procedure is very advantageous from the economic point of view, saving the expensive oxides materials and shortening the time of the total procedure.

The K_{eff} values were not calculated in the equilibrium state, therefore they varied with altered conditions, for instance when the equipment was changed. For two types of equipment we obtained values, listed in Table 2.

Table 2
Effective partition ratios of neodymium

Type of equipment	process number	$K_{eff\ beg.}$	$K_{eff\ mean}$	$\bar{K}_{eff\ mean}$
I	1	0.267	0.247	0.237
	2	0.241	0.227	
	3	0.288	0.256	
	4	0.286	0.246	
	5	0.220	0.211	
II	1	0.204	0.193	0.203
	2	0.228	0.214	
	3	0.218	0.206	
	4	0.215	0.199	
	5	0.216	0.203	

Conclusions

Investigations of the single crystal growth of the YAG type garnet by means of radioisotopic methods give valuable information concerning the process that take place during crystallization. It is possible to get the following data:

— the shape of the crystallization front
— homogeneity of neodymium distribution in the single crystal (changes in neodymium concentration along the single crystal axis have been revealed, as well as local concentration changes, connected with various types of segregation).
— neodymium content in the single crystal and in the residual material in the crucible, and from it, the effective partition ratio of neodymium.

A considerable influence of the thermal equipment type on neodymium segregation was found.

On the basis of the information obtained, the most suitable parameters can be chosen for crystal growing process and in consequence it is possible to obtain high quality single crystals and to diminish the cost of their production.

References

BRADSLEY, W., COCKAYNE, B.: J. Phys. Chem. Solids, Suppl. No 1, 109 (1967)
BRANDLE, C. D.: Report 1976, San Diego, California
COCKAYNE, B.: Phil. Mag. **12**, 943 (1965)
KESTIGIAN, M., HOLLOWAY, W. W.: in: Crystal Growth, Oxford 1967, pp. 451–456
MONCHAMP, R. R., BETT, E.: in: Crystal Growth, Oxford 1967, pp. 463–465
SMIRNOVA, E. P., KRAVCHUK, J. F., TIKHOMIROVA, E. J.: Cryst. Res. Technol. **20**, 1189 (1985)

(Received February 10, 1987)

Authors' addresses:
Dr. L. WALIŚ, Assist. prof. Dr. H. JASKÓLSKA
Institute of Nuclear Chemistry and Engineering
ul. Dorodna 16
03-195 Warsaw

Dr. Cz. JANUSZ
Institute of Electronic Materials Technology
ul. Wólczanska 133
01-919 Warsaw

Cryst. Res. Technol.	22	1987	9	1127–1137

An. König, Ax. König, H.-H. Emons

Sektion Chemie, Bergakademie Freiberg, Freiberg (DDR)

Crystallization Kinetics of $MgSO_4 \cdot 7\,H_2O$ at Higher Ranges of the Driving Force

By means of batch crystallization technique the crystallization kinetics of $MgSO_4 \cdot 7\,H_2O$ at 25 °C at higher ranges of the supersaturation from pure aqueous solutions was investigated. It was observed that the growth rate highly depends on the habit of crystals and on their ratio of length to width, respectively. This ratio is a function of the initial supersaturation of each crystal.

Die Kristallisationskinetik von $MgSO_4 \cdot 7\,H_2O$ aus reiner wäßriger Lösung bei 25 °C wurde im Bereich hoher Triebkräfte mit der Batch-Technik untersucht. Es wurde festgestellt, daß die Wachstumsgeschwindigkeit in starkem Maße von der Form der Kristalle, bzw. von ihrem Verhältnis Länge zu Dicke abhängt. Dieses Verhältnis ist eine Funktion der Ausgangsübersättigung eines jeden Kristalls.

1. Introduction

Crystallization is a multistage process where the steps of the desupersaturation — nucleation, crystal growth and secondary processes — are connected with each other by complicated feedback mechanisms and depend on each other. The process parameters such as supersaturation, hydrodynamics, temperature, but also the presence of impurities have a complex influence on each of these steps. Crystallization kinetics in the pure system as a function of the main contributing factors and under the same experimental conditions must be determined prior to investigation of the influence of a certain additive on the crystallization behaviour of a main component.

The most important prerequisite is given by experimental methods which ensure the possibility of keeping constant and varying systematically, respectively, the most important process parameters. The aim of the present paper is the determination of crystallization kinetics of $MgSO_4 \cdot 7\,H_2O$ at 25 °C in dependence on the driving force and with constancy of all remaining contributing parameters in the pure system. Special attention is paid to the investigation at high driving forces such as they can be used neither in the fluidized bed or in the MSMPR crystallizer nor in the single crystal experiment.

The previously given method of batch crystallization to obtain crystallization kinetic data by calorimetric measurement during crystallization and a crystal size analysis at the end of the experiment (König et al.) are especially suited to this purpose.

The characteristic feature of the system $MgSO_4 \cdot 7\,H_2O$, i.e. that the crystals change their shape with the size, fundamentally complicates the evaluation up to the kinetic equations of nucleation and growth compared with systems having a constant shape factor. Therefore, it is necessary to test whether the data obtained by the method of batch crystallization are in agreement with the data published in the literature — of course, only in the case of lower driving forces.

2. Experimental

Experimental equipment and the principle of evaluation of the batch crystallization experiments have previously been described in detail (KÖNIG et al.). The $MgSO_4 \cdot 7\,H_2O$ crystallization experiments were carried out at 25 °C at 3 different initial supersaturations. The dimensionless driving force

$$S = \ln\,(m_H/m_{H,sa}) \tag{1}$$

is used as a measure of the supersaturation state where m_H is the molality in *mol* $MgSO_4 \cdot 7\,H_2O$ per kg of H_2O. Calculation of the saturation molality in the temperature range of 25–45 °C with

$$m_{H,sa} = 1.712 + 0.1258\vartheta \tag{2}$$

is based upon the data given by TING and McCABE. The stirrer speed was maintained constant (500 min^{-1}) in all experiments. The used substances were twice recrystallized $MgSO_4 \cdot 7\,H_2O$ p.a. (VEB Laborchemie Apolda) and deionized water.

3. Results and discussions

3.1. Enthalpy and mass balance

The heat production curves which were found by calorimetric measurement at 3 different levels of the initial driving force from the chronomal temperature change during the crystallization experiments are shown in Figure 1.

The integral heat effect corresponds to the crystallization enthalpy of the amount of crystallizing $MgSO_4 \cdot 7\,H_2O$. The molar crystallization enthalpies obtained from a total of 25 crystallization experiments and the data published in the literature are summarized in Table 1. Therefrom it may be seen that no significant influence of the crystallizing amount of substance and the initial supersaturation, respectively, can be found with the accuracy of measurement which can be obtained by the method. Agreement with published data is satisfactory.

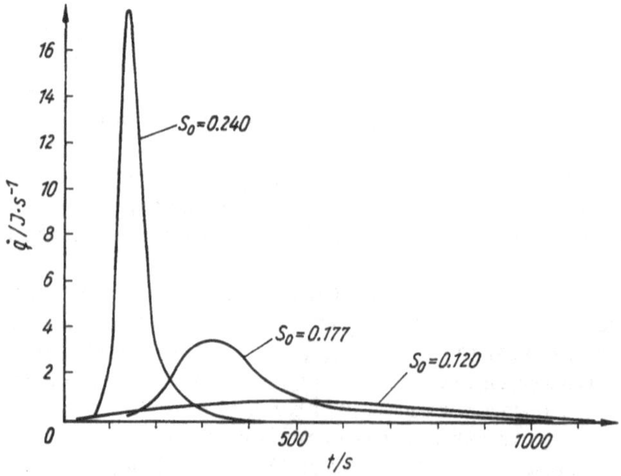

Fig. 1. Heat production curves as a function of the reaction time for the crystallization with 3 different initial supersaturations (in the ordinate read $-\dot{q}$)

Table 1
Crystallization enthalpy of $MgSO_4 \cdot 7\ H_2O$

$\vartheta/°C$	$\ln m_H/m_{H,sa}$	number of run	$\Delta_c H/kJ \cdot mol^{-1}$
25	0.120	7	24.18
25	0.177	2	23.14
25	0.240	16	23.52
25	mean value	25	23.66 ± 0.83 (±3.5%)

$\vartheta/°C$	$\Delta_c H/kJ \cdot mol^{-1}$	kind of investigation	author
25	22.5	direct measurement	Nývlt 1977
13	17.615	direct measurement	Perreu 1934
12	17.364	measured values of last heat of dissolution $\Delta_L H_{sa}$	Perreu 1934
19	20.292	measured values of last heat of dissolution $\Delta_L H_{sa}$	Perreu 1934
19	21.338	measured values of last heat of dissolution $\Delta_L H_{sa}$	Perreu 1935
20	19.874	calculated values of last heat of dissolution $\Delta_L H_{sa}$	Perreu 1934

Molality occurring at any time and corresponding supersaturation including a temperature correction as well as the actual magma density can be calculated from the heat production curves by means of mass balance according to the proportionality between heat and mass production.

Heat production curves graphically demonstrate differences in the reaction course, i.e. changes in the ratio of nucleation to crystal growth (Nývlt, Pekarek).

3.2. Crystal habit and size distribution

As shown by the photomicrographs in Figures 2a, b, c, different crystalline products exist depending on the initial driving force. The determined frequency distributions versus the characteristic crystal length are plotted in Figure 3. At the same time the figure shows a typical $MgSO_4 \cdot 7\ H_2O$ crystal with the characteristic crystal parameters (length L and width B) used in this work. B corresponds to a mean value of the two crystal expansions along the [110] and [1$\bar{1}$0] axial directions. The ratio of length to width with $v = L/B$ is considered as a measure of the habit altering of the crystals. As can be seen from Figure 4, v is the higher the longer are the crystals.

Of great importance is the fact that the crystal habit highly depends on the degree of supersaturation where nucleation occurs. As shown by the later described results, heavy nucleation occurs only up to a certain proportion of the starting supersaturation. The crystals shown in Figure 2a have mainly been formed in the supersaturation range of $0.120 > S > 0.010$ (corresponds to undercooling $\Delta T = 5$ K). As to be seen in Figure 4 the value v of these well-formed crystals reaches from 5 to 1.

When crystallization starts with double the initial driving force (nucleation range $0.240 > S > 0.040$), the largest first formed crystals show a ratio of nearly 20, whereas compact crystals are not formed. The crystals are the more needle-like the higher was their "birth supersaturation".

This fact is confirmed by numerous visual microscopic observations of nucleation. Contacting of a seed crystal leads to secondary nucleation only above a supercooling of about 3 K. Compact cubic to square crystals are formed. Crystals which are formed at the same time (i.e. with identical supersaturation) show a very identical shape. When undercooling was higher than about 10 K, a great number of very thin long needles are explosively formed which easily break owing to their fineness, i.e. they are formed with the lightest contact from a point of a seed crystal.

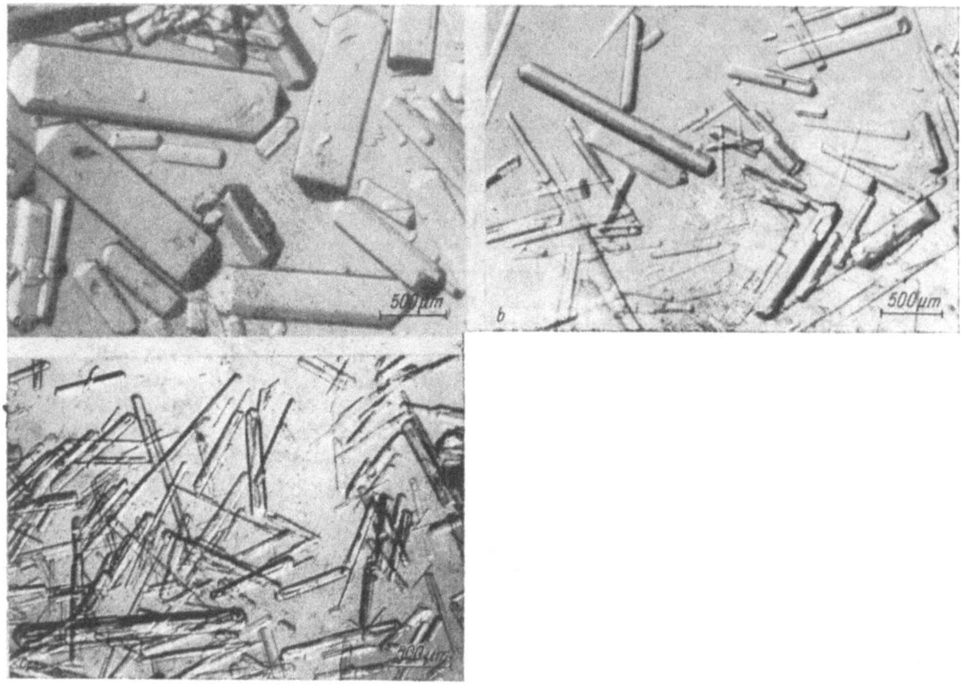

Fig. 2. Crystalline products at the end of the runs with starting supersaturation a) $S_0 = 0.120$; b) $S_0 = 0.177$ and c) $S_0 = 0.240$

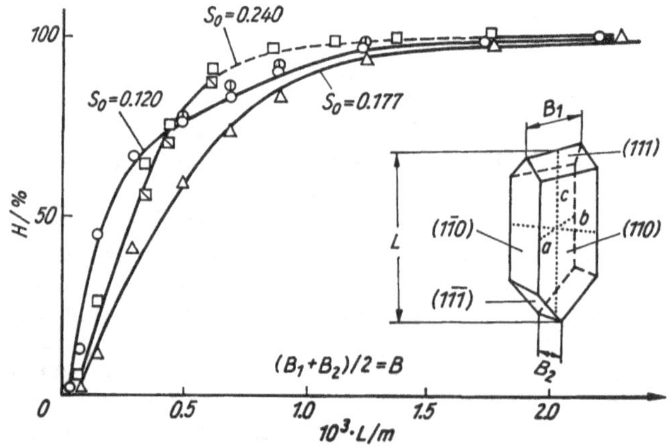

Fig. 3. Crystal size distribution in the form of the crystal size frequency for 3 different initial supersaturations. Typical $MgSO_4 \cdot 7\ H_2O$ crystal with crystallographic axis and the used characteristic dimensions L and B

These results and observations are in full agreement with the investigations performed by CLONTZ and MCCABE, of the secondary nucleation of $MgSO_4 \cdot 7\ H_2O$. For subcooling levels of 4 – 7 K the authors give a collision breeding mechanism where well-formed crystals were found. For higher supersaturations the growth mechanism of crystals is said to be changed. The ratio of v drastically increases. The term of "needle breeding" chosen by the authors is very convenient for this type of nucleation.

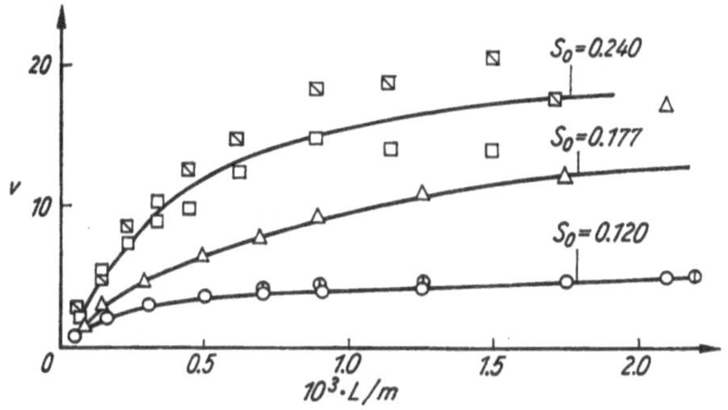

Fig. 4. Dependence of form factor $v = L/B$ on crystal length L for 3 different initial supersaturations

3.3. Linear crystal growth rate

As described in detail in another publication (König et al.), the mean crystal growth rate and the nucleation rate can be obtained as functions of the reaction time by combining of the mass balance in the chronomal course of the experiment with the crystal size balance at the end of the experiment. As the chronomal desupersaturation is calculated at the same time, the crystal growth rate can be given as a function of the driving force.

On calculation of the crystal growth rates in length direction $\dot{\bar{L}}$ and in the direction of the width $\dot{\bar{B}}$ it is assumed that

— the growth rate is the same for all crystals in one experiment

— and the ratio v of a crystal remains constant from a noticeable size (visible when magnified 100 times).

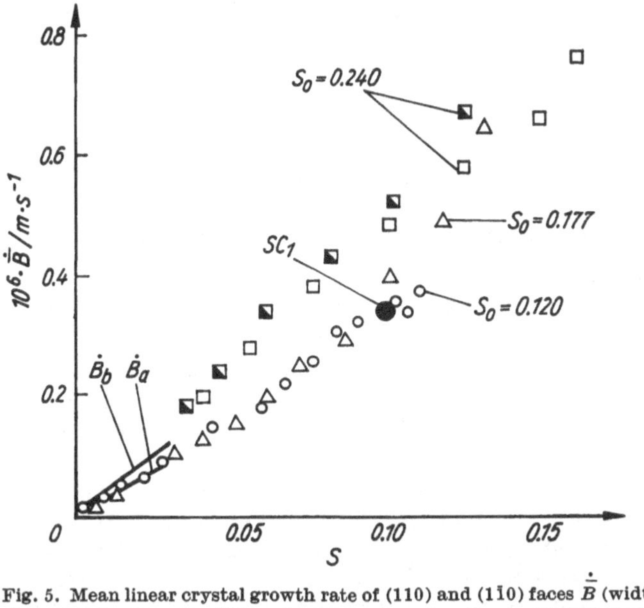

Fig. 5. Mean linear crystal growth rate of (110) and (1$\bar{1}$0) faces $\dot{\bar{B}}$ (width direction) as a function of the driving force for different initial supersaturation levels. SC_1 single crystal measurement at $w = 2 \cdot 10^{-2}$ m · s^{-1}; $\dot{B}_a$, $\dot{B}_b$ growth rate in direction of the crystallographic axis a and b from fluidized bed measurement published by Matuchova/Nývlt

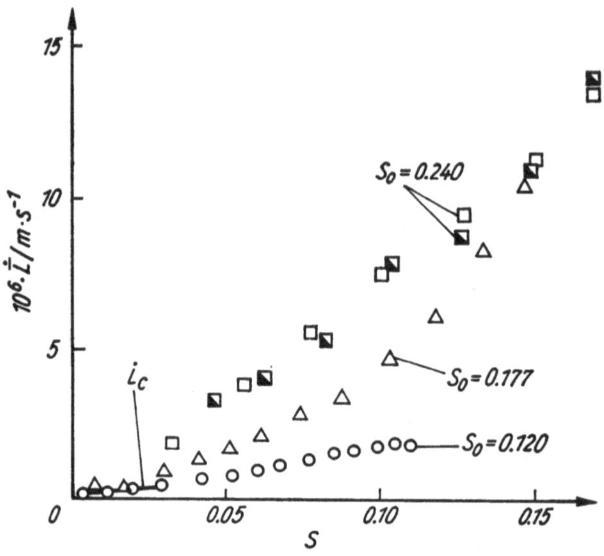

Fig. 6. Mean linear crystal growth rate in direction of the crystallographic c-axis (length direction) as a function of the driving force at 3 different initial supersaturation levels ($\dot{L}_C$ fluidized bed measurement of MATUCHOVA/NÝVLT)

The last-named assumption follows from the facts of the crystal habit dependence on the supersaturation described in 3.2. As shown in Figures 5 and 6, the mean crystal growth rates considerably differ in length and width direction in experiments with a different initial supersaturation level. Thus, growth for $S_0 = 0.24$ is four times more rapid than for $S_0 = 0.12$ with

$$\dot{\bar{L}}_{0.12} = 1.8 \cdot 10^{-5}\, S\,. \tag{3}$$

The thickness growth rates differ by the factor of 1.5.

As shown in 3.2., two kinds of crystal growth must be considered. The crystal growth rate obviously depends to a decisive extent on the ratio v of the crystals. Mainly in length direction they grow the faster the more they are needle-like. As each of the mean linear crystal growth rates $\dot{\bar{L}}$ and $\dot{\bar{B}}$ were calculated for each experiment for the crystalline product resulting at the end of the experiment with its characteristic shape and size distribution, the course deviating from the linearity of the functions $\dot{\bar{L}}$, $\dot{\bar{B}} = f(S)$ in the range of a higher supersaturation can be rated as a further proof of a growth dispersion relating to the crystal shape. (With a high supersaturation crystals having the highest v value are formed which, on the other hand, grow most rapidly).

This interpretation is supported in a very impressive manner by the results of the experiments starting with $S_0 = 1.77$ ($\triangleq$ 7.56 K supercooling). In this case the already mentioned range of the change in the mechanism of nucleation is exceeded in the course of the desupersaturation, i.e. at first needle-like crystals are formed, whereas later more compact crystals are formed to an increasing degree. Growth rate curves which highly deviate from the linearity clearly show that here it is a question of a synthesis of the growth functions of different types of crystals.

The phenomenon of size-dependent growth for $MgSO_4 \cdot 7\,H_2O$ is repeatedly treated in the literature (KARPINSKI; ROUSSEAU, PARKS). Our investigations clearly show that there is really no constant crystal growth rate for $MgSO_4 \cdot 7\,H_2O$ crystals. However, in contrast to the previous assumption of a size-dependent growth we take the opinion of a habit-dependent growth (dependent on the ratio of length to width) and the habit, on the other hand, is highly dependent on the "birth supersaturation" of each crystal. This assumption would explain the discrepancies between results of ROUSSEAU and PARKS, and of SIKDAR and RANDOLPH in different ranges of supersaturation, too.

On the other hand the good agreement of growth rates of compact crystals in our investigations ($S_0 = 0.12$) with results of MATUCHOVA and NÝVLT in fluidized bed experiments speaks well for it. They used crystals in a size range of $(1.8-2.3) \cdot 10^{-3}$ m having a ratio of v between 2.4 and 3.2. Their supersaturation range was $0.007 < S < 0.029$. As can be seen from Figures 5 and 6, their measured growth rates in all three directions (crystallographic axis a, b, c) are in a very good agreement with our results, although in our experiments the crystals had only an average size of less than a tenth ($0.17 \cdot 10^{-3}$ m at the 50% passage of the crystal size frequency distribution, Figure 3). Only a very small amount of crystals attains the order of magnitude of $\sim 2 \cdot 10^{-3}$ m. However, the ratios v show a very good agreement (our $v_{50\%} = 2.5$). Furthermore, a comparison with own single crystal experiments (replacement rate of the (110) face as a function of supersaturation and solution flow) in Figure 5 shows a good agreement with the "compact-crystal curve" (solution flow of $2 \cdot 10^{-2}\ \mathrm{m \cdot s^{-1}}$). The crystals used had a size of $(3-10) \cdot 10^{-3}$ m.

3.4. Nucleation rate

As earlier described (KÖNIG et al.), secondary nucleation in experiments with the respective initial supersaturation is started by inserting of a small seed crystal. Therefore, at the beginning of the crystallization process high supersaturation levels and low magma densities are present. In this range statements on the mechanism of nucleation can be made which was already discussed in 3.2.

In the range of low supersaturations (up to $S = 0.05$) and high magma densities conditions can be encountered such as they are found in MSMPR crystallizers or under industrial conditions. Accordingly, the results obtained there can be compared with those in quasistationary states and can be used for the calculation of technical crystallizers.

Such kinetics is usually correlated by semiempirical power law correlations for nucleation rates.

A nonlinear least-squares procedure was used to evaluate the data obtained in the mentioned experimental ranges of our batch studies. As the adaptation criterion of the fit the minimum of its relative standard deviation was chosen

$$s_{\mathrm{fr}} = \left\{\frac{\Sigma((\dot{N}_c - \dot{N}_e)/\dot{N}_c)^2}{n-k}\right\}^{0.5} \tag{4}$$

where n is the number of experimental data sets and k the number of numerical coefficients in the fitting equation (Report of IUPAC Commission, 1982).

The resulting power law kinetic expression is

$$\dot{N} = 1.45 \cdot 10^9\, S^{2.86} M_{\mathrm{D}}^{1.09}\,. \tag{5}$$

Considering the shape factors of compact $MgSO_4 \cdot 7\,H_2O$ crystals (with the ratio $v = 2$) follows the overall growth rate from the mean linear growth rate in length direction with

$$G = 0.60\dot{\bar{L}}_K \tag{6}$$

and with equ. (3)

$$G = 0.93 \cdot 10^{-5} S\,. \tag{7}$$

Consequently, in order to compare our result with the published data shown in Table 2 we can give the power law correlation by using the overall crystal growth rate. The published data are converted into the international system of units which was used by us. At the same time the table contains the experimental conditions.

In continuous experiments the driving force is difficult to measure. Therefore, the investigation range was estimated by us from our experimentally determined function $G = f(S)$ by means of the growth rate. An error in the order of magnitude seems to exist with the given supersaturation range of SIKDAR, RANDOLPH.

The comparison of our results with those of SIKDAR, RANDOLPH and ROUSSEAU, PARKS shows the same trend of nucleation rates depending on the driving force and magma density. The values of the kinetic parameters $k_{N,G}$, n and m are quite different and the absolute nucleation rates, too.

It is interesting to note here that each of the found coefficients of the power law fit highly depends on the adaptation criterion of the fit. Within the range of $s_{fr} < 0.52$ using the same experimental data set we can find the kinetic parameters in ranges of $1 \cdot 10^8$ to $1 \cdot 10^{11}$ for $k_{N,G}$, 2.76–2.96 for n and 0.30–1.69 for m. Important is the fact that a lot of different sets of kinetic parameters hold the fit with the same adaptation criterion. But we have always to consider the whole set, never the individual parameters. It is advisable to use always the same adaptation criterion for comparisons. The standard deviation of the fit (equ. (4)) recommended by the IUPAC is well suited to this purpose.

Because the experimental equipments which were used by the authors cited in Table 2 are substantially different from that of this study, it is not surprising that the nucleation rates are unequal.

The crystal size distribution in the three studies was measured in a different manner, the systems operate over significantly, different ranges of supersaturation and the most important fact, i.e. the hydrodynamic conditions are very different, too. Accordingly, the lower nucleation rates have to be taken into consideration in using kinetic data from batch studies.

4. Conclusions

Prior to investigations of the influence of impurities the crystallization kinetics of $MgSO_4 \cdot 7\,H_2O$ was investigated in dependence on the driving force and with constancy of all remaining crystallization parameters in the pure system at a temperature of 25 °C. Results of the crystal growth rate in length and width direction of the $MgSO_4 \cdot 7\,H_2O$ crystals were obtained by the method of batch crystallization which are in a very good agreement with published data of other experimental methods with lower supersaturation. Crystal nucleation rates in the range of high magma densities can also be compared with those which are obtained in quasistationary states.

The results show that in contrast to the size-dependent growth repeatedly cited in the literature the growth rate depends more likely on the ratio of length to width of the crystals, which, on the other hand, is a function of the initial supersaturation of

Table 2

Nucleation rates of $MgSO_4 \cdot 7\ H_2O$ and experimental conditions

experimental method	$\dot{N} = K_{N,G} G^n \cdot M_D^m$ $m^{-3}\ s^{-1}$			$\dot{N}$[1]	$G \cdot 10^8$	S	M_D	ϑ	author
	$K_{N,G}$	n	m	$m^{-3}\ s^{-1}$	$m \cdot s^{-1}$	$\ln (m_H/m_{H,sa})$	$kg \cdot m^{-3}$	°C	
MSMPR	$9.65 \cdot 10^{15}$	1.24	0.67	$1.1 \cdot 10^8$	2.6–7.1	0.0266–0.064 [2]) 0.0028–0.0076	—	18–26	SIKDAR/ RANDOLPH
perfectly mixed cont. crystallizer	$3.51 \cdot 10^9$	0.51	0.68	$9.1 \cdot 10^6$	1.3–5.5	—	17–128	26–34	ROUSSEAU/ PARKS
	[3]) $6.8 \cdot 10^{11}$	0.85	0.80	$9.1 \cdot 10^6$		[2]) 0.0014–0.0059			
batch crystallizer	[4]) $3.56 \cdot 10^{23}$	2.86	1.09	$2.7 \cdot 10^4$	5.6–58.6	0.006–0.066	58–103	25	present work
	[5]) $5.62 \cdot 10^{20}$	2.38	1.42	$5.6 \cdot 10^5$	27–202	0.0026–0.196	26–150	25	

[1]) absolute value for $M_D = 50$ kg m^{-3} and $G = 4.65 \cdot 10^{-8}$ $m \cdot s^{-1}$ or $S = 0.005$
[2]) calculated with $S = 1.0753 \cdot 10^5\ G_K$ from equ. (7)
[3]) nonlinear least-squares procedure using S_{fr} (equ. (4)) as an adaption criterion of the fit
[4]) compact crystals
[5]) needle-like crystals ($v = 20$; $G = 0.14\ \dot{\bar{L}}_N$; $G = 1.03 \cdot 10^{-5} S$)

each crystal. Consequently, a crystal growth dispersion results from the supersaturation where the individual crystals have been formed. This knowledge is mainly found by comparison of the crystallization experiments which are carried out with a very high initial supersaturation ($\Delta T = 10$ K) with those being performed with a lower driving force.

The authors wish to thank Dr. J. Nývlt for stimulating and helpful discussions.

Symbols

B	crystal width	m
$\dot{\bar{B}}$	mean linear crystal growth rate of $(\bar{1}10)$ and (110) faces	$m \cdot s^{-1}$
G	overall crystal growth rate	$m \cdot s^{-1}$
H	relative frequency	%
$\Delta_c H$	molar crystallization enthalpy	$kJ \cdot mol^{-1}$
k_L	kinetic coefficient of crystal growth	$m \cdot s^{-1}$
$k_{N,S}$; $k_{N,G}$	kinetic coefficient of nucleation	$\# m^{-3} \cdot s^{-1}$
L	characteristic length of crystals	m
$\dot{\bar{L}}$	mean linear crystal growth rate in direction of L	$m \cdot s^{-1}$
m_H	molality	mol hydrate salt/kg H_2O
M_D	magma density	$kg \cdot m^{-3}$
$\dot{N}$	nucleation rate	$\# m^{-3} \cdot s^{-1}$
$\dot{q}$	heat production	$J \cdot s^{-1}$
S	driving force $S = \ln (m_H/m_{H,sa})$	
s_{fr}	relative standard deviation of the fit	
v	ratio $v = L/B$	
w	solution flow velocity	$m \cdot s^{-1}$
ϑ	temperature	°C
Δ	difference	

Indices

K	compact, well-formed
N	needle-like
sa	saturated state
0	initial state

References

Clontz, A., McCabe, L.: Chem. Engng. Progr., Symp. Ser. **67** (No. 110) 6 (1971)
IUPAC Commission 1.2 on Thermodynamics: J. Chem. Engng. Data **27**, 221 (1982)
Karpinski, P. H.: Kristall u. Technik **15**, 507 (1980)
König, Ax., König, An., Emons, H.-H.: Chem. Techn. **37**, 283 (1985)
Matuchova, M., Nývlt, J.: Chem. prum. **25/50**, 463 (1975)

NÝVLT, J.: Industrial Crystallization, Amsterdam—Oxford—New York—Tokyo 1977
NÝVLT, J., PEKAREK, V.: Z. phys. Chem. (Frankfurt/Main) **128**, 197 (1981)
PERREU, J.: C. r. Acad. Sci. Paris **198**, 172, 1410, 1767 (1934); **199**, 48 (1934); **200**, 1030, 1588 (1935) in RYCHLY, R., NÝVLT, J.: Kristall u. Technik **9**, (1974)
ROUSSEAU, R., PARKS, R.: Ind. Engng. Chem. Fundam. **20**, 71 (1981)
SIKDAR, S., RANDOLPH, A.: AlChE **22**, 110 (1976) in GARSIDE, J., SHAH, M. B.: Ind. Eng. Chem. Process Des. Dev. **19**, 509 (1980)
TING, H. H., MCCABE, W. L.: Ind. Engng. Chem. **26**, 1201, 1207 (1934)
TREIVUS, E. B.: J. fis. chim. 1088 (1979)

(Received February 18, 1987)

Authors' addresses:

Dr. An. KÖNIG, Prof. Dr. Drs. h. c. H.-H. EMONS
Bergakademie Freiberg, Sektion Chemie
Leipziger Straße
Freiberg
DDR-9200

Doz. Dr. Ax. KÖNIG
Pädagogische Hochschule Erfurt/Mühlhausen
Sektion Chemie/Biologie
Schillerweg
Mühlhausen
DDR-5700

Book Review

M. Fujimoto (Editor)

Gallium Arsenide and Related Compounds 1985

Institute of Physics Conference Series 79
Proceedings of the Twelfth International Symposium on Gallium Arsenide and Related Compounds held in Karuizawa, Japan, 23—26 September 1985

Adam Hilger Ltd. Bristol and Boston 1986, 754 Seiten, zahlreiche Abbildungen und Tabellen, Autorenindex, £ 45.00, ISBN 0-85498-170-5

Je höher der Entwicklungsstand in der Silizium-Bauelementetechnologie fortschreitet, um so deutlicher werden Grenzen sichtbar, die durch die Grundeigenschaften dieses Materials gegeben sind. Diese Tatsache spiegelt sich in dem Konferenzband besonders wider, vor allem, was die Entwicklungen auf dem Gebiet der Höchstgeschwindigkeitsbauelemente angeht. Der Leser erhält einen Überblick über Arbeiten, die etwa in den Jahren 1983 bis 84 in den Forschungslaboratorien entstanden und von denen einige auch heute von starkem Interesse sind. Was die Palette der untersuchten Materialien betrifft, so hat in der letzten Zeit innerhalb der $A^{III}B^{V}$-Verbindungen eine gewisse Profilierung stattgefunden. Antimonide spielen kaum eine Rolle, GaP wird für derzeit stehende Anforderungen bei LED's beherrscht.

Das Buch enthält 13 Kapitel. Das Spektrum der Themen reicht von der Kristallzüchtung bis zu Bauelementestrukturen. Schwerpunkt auf dem Substratsektor ist die Entwicklung von undotiertem halbisolierendem GaAs und dotiertem InP mit möglichst geringer Versetzungsdichte. Bei den Epitaxieverfahren ist die bewährte, aber für dünnste Schichten kaum einsetzbare Flüssigphasenepitaxie gegenüber den verschiedenen Varianten der OMVPE- und der MBE-Verfahren in den Hintergrund gerückt, nachdem man gelernt hat, zwar mit höchstem technischem Einsatz, aber mit Erfolg, Anfangsprobleme der Reinheit und der Eigenstörstellen zu lösen. Gerade bei der MBE ist es gelungen, neben vertikaler Strukturierung auch eine laterale in-situ durchzuführen. Superlattices, Quantum wells, Modulationsdotierung und impurity induced disordering stehen gegenwärtig im Brennpunkt des Interesses. Aus den Beiträgen wird deutlich, daß das Schwergewicht der Charakterisierung anwendungsorientiert ist. Es werden die Eigenschaften untersucht, die Bauelementerelevanz haben. Damit werden neue Anforderungen an die Meßtechnik gestellt.

Von den insgesamt fast 800 Seiten des Bandes ist die Hälfte Problemen der Prozeßtechnologie bzw. der Entwicklung diskreter Bauelemente und integrationsfähiger Strukturen gewidmet, so auch der Kombination von $A^{III}B^{V}$-Systemen auf Si-Substraten. 16 von 146 Arbeiten befassen sich mit Lasern unter dem Aspekt der Lichtleiternachrichtenübertragung.

Das Buch spiegelt interessante Entwicklungstendenzen der letzten Jahre wider und zeigt, mit welchem Aufwand das Forschungsgebiet betrieben werden muß. Es ist für Materialentwickler genauso ergiebig, wie für Physiker und Bauelemente-Fachleute. In Inhalt, Form und Qualität fügt es sich ohne „Fehlpassung" in die in den letzten 10 Jahren zur gleichen Thematik erschienenen Konferenzbände dieser Serie ein.

G. Kühn

Cryst. Res. Technol.	22	1987	9	1139–1144

Irina Nicoară, D. Nicoară, O. F. G. Aczel

Department of Physics, University of Timişoara, Timisoara, Romania

Crystal Growth in an Improved Bridgman Variable Shape Graphite Furnace

An improved Bridgman-type crystal growth system is presented. By using a new type of heaters, fluoride single crystals with a diameter up to 50 mm and 150 mm length have been obtained. The dislocation density distribution has been examined in different growth conditions.

Es wird ein verbessertes Bridgman-System für die Züchtung von Einkristallen bekanntgegeben. Mit diesem System wurden Calciumfluorid-Einkristalle der Länge 150 mm und mit Durchmessern bis zu 50 mm erhalten unter Verwendung einer neuen Art Heizer verwendet wurde. Die Dichte und Verteilung der Versetzungen wurde für verschiedene Wachstumsbedingungen untersucht.

1. Introduction

One of the most common techniques for single crystal growth from melt is a variant of the original Bridgman-Stockbarger method (Bridgman; Stockbarger). Their main advantages consist in: the possibility of growing crystals with a constant diameter, a small specific consumption of electric power, because of a better containment of the generated heat and the relative facility to achieve complete automation of the process. The requirement of a high temperature gradient at the interface of crystallization, which is incompatible with the low temperature gradient needed to succesfully cool down the crystal is a major disadvantage of both methods. In the absence of comprehensive studies of the relationship between controllable experimental parameters and the associated crystal growth characteristics, process optimization has been so far restricted to empirical approaches.

Lately, significant experimental (El-Mahallawy, Farag; Clyne; Wang, Witt) and theoretical (Chang, Wilcox; Fu, Wilcox; Clyne; Naumann; Jasinski, Rohsenow; Huang et al.) works have been reported for the Bridgman-Stockbarger configuration, providing thermal boundary conditions that lead amenably to a meaningful thermal analysis. The results of heat transfer analyses have mainly been confirmed by experimental investigations. The purposes of such theoretical investigations were to provide basic criteria for the establishment of such conditions that will lead to improved chemical and crystalline perfection during the conventional Bridgman growth, to establish the applicability of the results obtained from theoretical heat transfer studies and to offer an experimental basis for the design of an improved Bridgman-type growth system.

The aim of the present paper is to describe a new type of variously shaped graphite heaters by means of which an optimized temperature profile for Bridgman crystal growth is obtained, heaters that eliminate the need of two separate heating chambers. We also report the optimum conditions for the growth of good quality CaF_2-type crystals by studying their dislocation etch pit distribution.

2. Description and characteristics of the heater

An important factor controlling the quality and perfection of any crystal grown from melt is the temperature gradient at growth interface. An interface formed of a concave solid permits misoriented regions nucleated at the external surface to grow into the bulk of the crystal. Supercooling at interface will produce dendritic growth and in doped materials — microsegregation of the dopant. The temperature gradient in the grown crystal affects both the density and the distribution of dislocations and as well as its internal strain.

Taking into account the analysis of the equilibrium interface shape, a concave isotherm toward the liquid should be realized in order to eliminate the formation of facets and dendritic growth. As a consequence the crystallization of the melt must take place in a high temperature gradient zone. This condition is essential in order to obtain single crystals of large dimensions.

The heat flow characteristics, which greatly influence the thermal profile at the interface, are often determined by the thermal properties of the material being frozen and its container, the relative geometries of both the sample and the furnace, as well as the rate of freezing (JASINSKI, ROHSENOW). The major effect of the sample is to decrease the thermal gradient, especially in the case of large crystals, the change being further dependent on the freezing rate and the length of the sample.

FU and WILCOX investigated a three-zone furnace in which the hot and cold zones were separated by a short adiabatic zone. Computer modelling using a finite difference scheme was employed, which allowed unequal heat-transfer coefficients to be considered in the hot and cold zones. It was shown that the use of an adiabatic zone significantly reduced the curvature of the isotherms between the hot and cold zones and allowed a much better control over the shape of the melt interface. This benefit, however, is at the expense of some loss in the temperature gradient. In the design of vertical Bridgman type directional solidification experiments, it is important to predict the thermal profiles in the specimen. The position and shape of the solidification interface and the axial thermal gradients on each side of the interface are of particular interest for the control of the process.

Similary, by taking into consideration the JASINSKI and ROHSENOW analysis, which recommends a certain proportion between the lengths of the three zones — hot, cold and gradient zone — referring to the diameter and the length of the crystal, several types of graphite heaters were designed (NICOARĂ; NICOARĂ et al. 1982; NICOARĂ et al. 1985), with their characteristics satisfying the above mentioned demands.

In the present paper we propose a new type of graphite heater that eliminates the need of using two heating chambers, independently controlled and with an intermediate diafragm, recommended in the Bridgman technique; this heater can produce temperatures of up to 2200 °C in vacuum of 10^{-4}—10^{-5} torr. The model contains an adiabatic zone and a booster heater zone that can be used to increase the thermal gradient near the solidification interface.

The general view of the crystal growth apparatus is shown in Figure 1. With this apparatus fluoride single crystals up to 50 mm diameter, of good optical quality and dislocation densities of 10^3—10^4 cm^{-2} can be obtained.

Figure 2 illustrates the longitudinal section of two types of heaters and the temperature distribution along them. Built of graphite, the heater produces an almost constant temperature zone (the hot zone B—C) necessary for the melting of the substance and a high gradient temperature zone (D—E). The C zone (the booster heater) is conceived as a meander-type resistance, by means of which an overheating is obtained in the lower level of the upper zone. The D zone is inserted for the improve-

Fig. 1. Photograph of the apparatus: 1 – hot zone (its wall is water-cooled), 2 – water-cooled flakes, 3 – molybdenum and graphite screens, graphite heater and crucible, 4 – keyboard, 5 – lower part of the apparatus with water-cooled power lines, 6 – pipe leading to vacuum pump

ment of heat transfer towards the upper part of the heater, the E zone, which appears like a wall thickening, has the role of an adiabatic zone and leads to the obtaining of the high temperature gradient. The F zone — the cold one — is used as the lower cooling chamber. The electrical and thermal characteristics of the furnace are the following: for a working temperature of 1400 °C the electrical power is 4.2 kW, with a voltage supply of 16 V, 50 Hz; for 2200 °C the electrical power used is about 8 kW. An electronic regulation and automation system assures the thermal stability of the apparatus. In order to avoid thermal loss, the graphite heater is surrounded by a set of concentric screens made of molybdenum, graphite, and stainless steel. In order to obtain small-diameter crystals (up to 10 mm), the zone C can be eliminated, as the necessary temperature gradient can be obtained by means of heater wall thickness adjustments (Fig. 2).

3. Results and discussion

By using the previously described heaters, CaF_2, BaF_2, etc., single crystals up to 50 mm diameter and 150 mm length can be obtained, whereas by means of a special crucible, oriented single crystals can also be produced. The spectral pure graphite crucible consists of a body with an extension for seed in its lower side, whereas the upper part has a container in which an extra quantity of material is kept, inducing the growth of a crystal of the desired length (Nicoară et al. 1985). The crucible is centered within the heater with the seed about 1 cm below the maximum temperature point, so that only a part of the seed will melt. As a result of the growth process, a single crystal with the same orientation as the seed is obtained (Fig. 3, A). In the

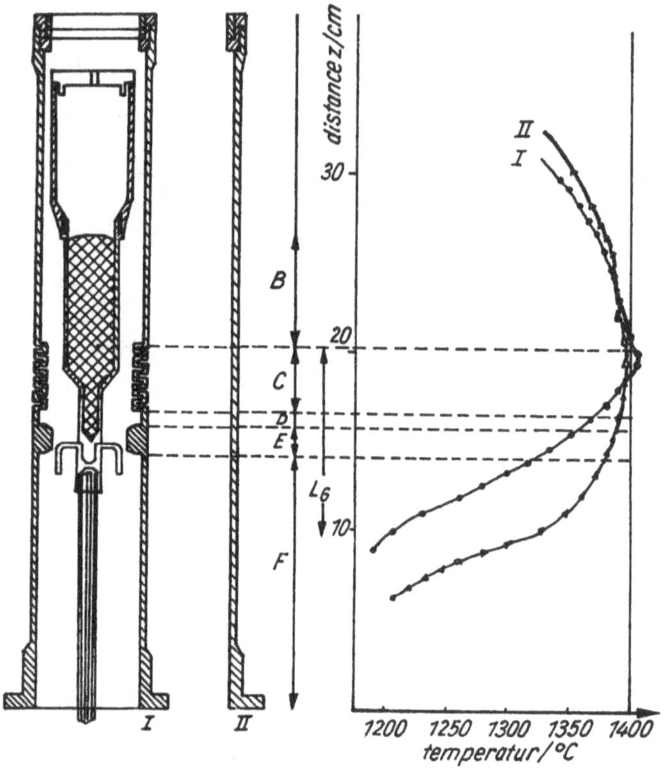

Fig. 2. Longitudinal sections of both types of heaters and the temperature distribution along them

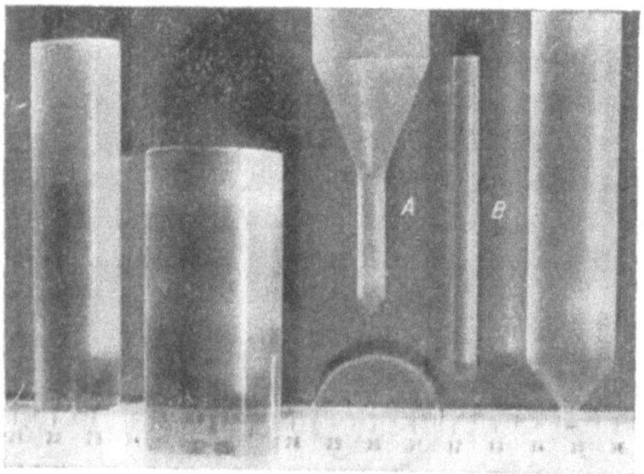

Fig. 3. Photograph of CaF_2 crystals of different diameters; crystal with seed (A)

lower part of the crucible a molybdenum radiator is attached in order to keep part of the seed unmelted.

Due to the remarkable optical properties of fluoride crystals, the problem of obtaining them in large quantities has often been raised. With a multiple type crucible (NICOARĂ et al. 1985) several up to 12 mm diameter crystals (Fig. 3, B) can be obtained during a single growth process.

The crystals are clear, colourless and have good cleavage planes. For cleaved 1 mm thick slides the transmission coefficient is 98% in both visible and infrared where it is up to 9 μm for CaF_2 and 12 μm for BaF_2. The quality of the crystals was studied by examining their dislocation distribution. For this purpose the crystals were cut into slices upon the cleavage surface (111) and their surfaces were etched to show up dislocation pits. Fresh cleavages were subjected to etching in aqueous solution using 2 N HCl. Experiments were carried out for different periods of etching at 30 °C in a temperature-controlled water bath for maintaining the temperature constant. They were then cleaned and examined under an optical microscope. The density of etch pits revealed was found to be independent on exposure time to the etchant, being a characteristic of the crystal itself rather than of the etch treatment (NICOARĂ et al. 1986).

The formation, multiplication and high mobility of the dislocations in ionic crystals lead not only to high densities of individual dislocations, but also cause arrangements of dislocations into well-developed grain sub-boundaries, which are stable and nearly immobile in contrast to individual dislocations. Subboundaries cannot be easily removed by annealing. The dislocation density along the sample and for a given cross section in radial direction as well have been studied. Some sub-boundaries can appear at the beginning of the growth. If their network presents merely a few meshes and if no new dislocations are generated during the further growth, these sub-boundaries grow out gradually on the face of the crystal and their number decreases at its upper end.

The studies of the etched cross-section of the crystal taken at different levels indicates the following: the density of pits and sub-boundaries is quite low and fairly uniform through the volume (Fig. 4a, b) in up to 12 mm diameter crystals which were

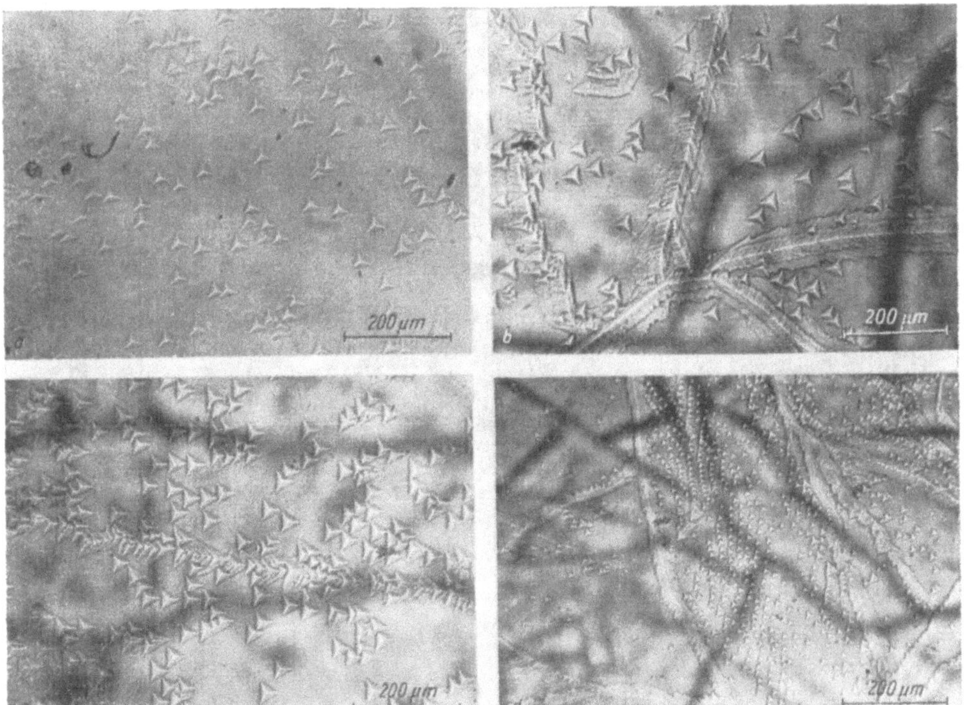

Fig. 4. Dislocation distribution on the (111) cleavage surface: lower part (a) and upper part (b) of the crystal; crystal obtained at high pulling speed (c) and (d). Etch conditions: 2 N HCl at 30 °C, (a), (b) and (c) for 3 hours and (d) for 1 hour

grown in a I type heater with a 6 mm/h speed and the temperature gradient of 7 °C/cm. The dislocation density is about 10^3—10^4 cm^{-2}. For a higher pulling speed the dislocation denisty increases rapidly to a much higher value at the upper end of the ingot (Fig. 4c, d).

For obtaining crystals with diameters larger than 12 mm, a meander type heater must be used in order to ensure the necessaıy temperature gradient at the melt-solid interface; the temperature gradient is about 20 °C/cm and the pulling speed 4 mm/h (Fig. 2, I).

An interesting feature is shown in the cross-section of the crystals that are grown in multiple crucibles: the dislocation density is greater in that part of the crystal which is oriented towards the axis support of the crucible. This fact is explained by the existence of an inadequate radial temperature gradient, the crucible being asymmetrically placed in the termic field of the furnace, one of its parts near the hot wall. This disadvantage is eliminated by attaching a reflective molybdenum cylindrical screen to the axis, which ensures the existence of a uniform thermic field for the crucible assembly.

By establishing the A—E heater zone dimensions and bearing in mind the thermal characteristics of both the crystal and the crucible, an improved Bridgman-type system can be realized, in which the crucible lowering in the heater will not unfavourably modify the axial temperature gradient, especially at the melt-solid interface, resulting in good quality crystals.

References

BRIDGMAN, P. W.: Proc. Amer. Acad. Arts Sci. **60**, 305 (1925)
CHANG, C. E., WILCOX, W. R.: J. Cryst. Growth **21**, 135 (1974)
CLYNE, T. W.: J. Cryst. Growth **50**, 684 (1980)
CLYNE, T. W.: J. Cryst. Growth **50**, 691 (1980)
EL-MAHALLAWY, N. A., FARAG, M. M.: J. Cryst. Growth **44**, 251 (1978)
FU, T. W., WILCOX, W. R.: J. Cryst. Growth **48**, 416 (1980)
HUANG, C. E., ELWELL, D., FEIGELSON, R. S.: J. Cryst. Growth **64**, 441 (1983)
JASINSKI, T., ROHSENOW, W. M., WITT, A. F.: J. Cryst. Growth **61**, 339 (1983)
NAUMANN R. I.: J. Cryst. Growth **58** 554 (1982)
NICOARĂ, D.: Patent R.S.R. Nr. 62842 (1972)
NICOARĂ, I., ACZEL, O. F. G., NICOARĂ, D., SCHLETT, Z.: Cryst. Res. Technol. **21**, 647 (1986)
NICOARĂ, D., NICOARĂ, I., SCHLETT, Z.: Patent R.S.R. Nr. 88497 (1985)
NICOARĂ, D., NICOARĂ, I., SCHLETT, Z.: Patent R.S.R. Nr. 85993 (1985)
NICOARĂ, D., SCHLETT, Z., NICOARĂ, I.: Patent R.S.R. Nr. 82663 (1982)
STOCKBARGER, D. C.: Rev. Sci Instrum. **7**, 133 (1936)
WANG, C. A., WITT, A. F., CARRUTHERS, I. R.: J. Cryst. Growth **66**, 684 (1984)

(Received February 20, 1987)

Authors' address:

Irina NICOARĂ, D. NICOARĂ, Otto ACZEL,
Universitatea din Timişoara, Catedra de Fizică
Bd. V. Pîrvan No 4
1900 Timişoara Romania

Cryst. Res. Technol.	22	1987	9	1145–1151

H. GÖRLS, D. WALTHER, J. SIELER

Friedrich-Schiller-Universität Jena, Sektionen Physik and Chemie, and
Karl-Marx-Universität Leipzig, Sektion Chemie

Crystal and Molecular Structure of Bis(glyoxal-bis-tert.butylimine)nickel(o)

The crystal structure determination of the title compound was undertaken to determine the coordination geometry of nickel atom. Bis(glyoxal-bis-tert.butylimine)nickel(0) ($C_{20}H_{40}NiN_4$) crystallizes in the space group $Pna2_1$ with $Z = 4$, $a = 18.546(6)$, $b = 9.331(8)$, and $c = 13.505(8)$ Å. The structure was solved by the heavy atom method and refined to $R = 0.065$. The molecule has four-coordinate nickel atom. The coordination geometry about the nickel atom is tetragonal.

Die Kristallstrukturanalyse der Titelverbindung wurde durchgeführt, um die Koordinationsgeometrie des Nickels zu bestimmen. Bis(glyoxal-bis-tert.butylimin) nickel(0) ($C_{20}H_{40}NiN_4$) kristallisiert in der Raumgruppe $Pna2_1$ mit $Z = 4$, $a = 18.546(6)$, $b = 9.331(8)$ und $c = 13.505(8)$ Å. Die Struktur wurde durch die Schweratommethode geöst und bis zum R-Wert von 0.065 verfeinert. Das Molekül besitzt ein vierfach-koordinieres Nickelatom. Die Koordinationsgeometrie des Nickels ist tetragonal.

1. Introduction

Glyoxal-bis-tert.butylimine is among the great group of "diazadiene" (dad). On the strength of electronic and stereoscopic features diazadiene can stabilize metals in a low oxidation stage. The first $(dad)_2$ Ni(0)-complex was prepared in 1966 from BALCH and HOLM. In the middle of the seventies two groups (WALTHER et al. 1974, 1975; TOM DIECK et al. 1975) prepared independent on each other some bis-diazadiene complexes with aromatic as well as aliphatic N-substitutions. By reason electronic and stereoscopic influences in function of N-substitution was expected a different coordination about the nickel atom (WALTHER 1977). For secondary and tertiary alkyl substitutes was awaited exclusive a tetragonal geometry. A distortion to a planar geometry was expected for aromatic substitutes. 1981 the crystal structures were determined from two compounds of this complex class — an aromatic and a secondary alkyl substitute $(dad)_2$ Ni(0)-complex (TOM DIECK et al. 1981). The results confirmed the correctness for this assumption. With the crystal structure of the title compound the experimental proof should be offered for a representative of a tertiary N-substitute $(dad)_2$ Ni(0)-complex too.

Besides in the last time it was found that $(dad)_2$ Ni(0)-complexes act as controlling ligands for the very selective cyclotetramerization of functionalized alkynes like $HC{\equiv}C{-}CH_2OH$ and $HC{\equiv}C{-}C({=}O){-}OR$ (TOM DIECK et al. 1985). Therefore it is interesting to know the structure too.

2. Experimental

For the X-ray diffraction a violet crystal of octahedral shape was used. Oscillation and Weissenberg photographs showed that the crystal is orthorhombic. A systematic extinction of diffractions of the $0kl$ type for $h + k = 2n + 1$ and of the $h0l$ type for $h = 2n + 1$ indicated two possible space groups of the symmetry $Pna2_1$ or Pnma. The lattice parameters were refined on Syntex P 2_1 diffractometer by the least-squares method from positional angles of 15 selected reflections using CuKα radiation. The integral intensities were measured on the same diffractometer using graphite monochromatized CuKα radiation and $\Theta - 2\Theta$ scan technique in the interval $0 \leqq 2\Theta \leqq 100°$. No significant fluctuations were observed in intensities of two standard reflections measured in an interval of 98 measured reflections. Intensities of all reflections were corrected for Lorentz and polarization factors. Empirical absorption correction (Taper) was made as well. From 1273 measured reflections, 1142 with $I \geqq 3\sigma(I)$ were regarded as observed.

A summary of the crystallographic data are listed in Table 1.

Table 1

Crystal data

Crystal morphology	violet octahedron
Chemical formula	$C_{20}H_{40}NiN_4$
Molecular weight	395.3 g/mol
Crystal system	orthorhombic
Space group	$Pna2_1$
Cell constants	$a = 18.546(6)$ Å
	$b = 9.331(8)$ Å
	$c = 13.505(8)$ Å
Volume	2337.1(3) Å^3
Number of formula units Z	4
Density (calculated) D_c	1.12 g/mol
Absorption coefficient μ(CuKα)	11.31 cm^{-1}
unique data measured	1273
Observed data with $I \geqq 3\sigma(I)$	1142
R	0.061
R_w	0.065
Rest electron density	0.435 e/Å^3

3. Structure analysis

The structure was solved by the heavy atom method. The three-dimensional Patterson synthesis calculated from all observed reflections showed position of the nickel atom and unambiguously determined the space group as $Pna2_1$. The positions of other non-hydrogen atoms were determined from successive Fourier syntheses of electron density. The positional and isotropic temperature parameters were refined by the full matrix least squares for all observed reflections to $R = 0.112$. After generation of the H atom positions from assumed geometries the refinement of the nonhydrogen coordinates with anisotropic vibrational parameters led to $R = 0.074$. In the final stage of the least-squares procedur H positions were refined in isotropic mode. This converged with a final $R = 0.065$. The resulting atomic parameters are listed in Table 2. All calculations were carried out on a ESER 1054 computer using the SHELX-76 program (Sheldrick).

Table 2
Atomic coordinates with e.s.d.

Atom	x/a	y/b	z/c	Atom	x/a	y/b	z/c
Ni1	0.1733(1)	0.2031(1)	0.2500(4)	H1-6-1	0.171(2)	0.092(2)	0.472(2)
N1-1	0.115(1)	0.302(1)	0.344(1)	H1-6-2	0.179(2)	0.209(3)	0.574(2)
N1-2	0.113(1)	0.303(1)	0.155(1)	H1-6-3	0.233(2)	0.235(3)	0.469(2)
C1-1	0.063(1)	0.388(2)	0.305(2)	H1-8-1	0.030(2)	0.149(3)	0.033(2)
C1-2	0.064(1)	0.382(2)	0.201(1)	H1-8-2	0.046(2)	0.249(3)	-0.075(2)
C1-3	0.121(1)	0.300(1)	0.456(1)	H1-8-3	0.004(2)	0.331(3)	0.028(2)
C1-4	0.049(1)	0.222(4)	0.499(2)	H1-9-1	0.181(2)	0.486(3)	0.026(2)
C1-5	0.136(2)	0.446(3)	0.493(2)	H1-9-2	0.088(1)	0.523(2)	0.023(2)
C1-6	0.181(2)	0.202(3)	0.495(2)	H1-9-3	0.130(1)	0.442(2)	-0.079(2)
C1-7	0.113(1)	0.296(2)	0.045(1)	H1-10-1	0.185(1)	0.182(2)	-0.055(2)
C1-8	0.044(2)	0.253(3)	0.005(2)	H1-10-2	0.170(1)	0.094(2)	0.058(2)
C1-9	0.129(1)	0.449(2)	0.001(2)	H1-10-3	0.225(1)	0.248(2)	0.054(2)
C1-10	0.178(1)	0.196(2)	0.024(2)	H2-1-1	0.364(1)	0.082(1)	0.243(2)
N2-1	0.277(1)	0.208(1)	0.252(2)	H2-2-1	0.262(1)	-0.149(1)	0.257(2)
N2-2	0.187(1)	-0.001(1)	0.254(2)	H2-4-1	0.346(2)	0.342(3)	0.406(2)
C2-1	0.306(1)	0.076(1)	0.249(2)	H2-4-2	0.414(2)	0.428(3)	0.338(2)
C2-2	0.257(1)	-0.034(1)	0.255(2)	H2-4-3	0.412(2)	0.239(3)	0.345(2)
C2-3	0.327(1)	0.334(1)	0.249(2)	H2-5-1	0.312(1)	0.564(1)	0.254(2)
C2-4	0.380(2)	0.335(3)	0.341(2)	H2-5-2	0.256(1)	0.461(1)	0.331(2)
C2-5	0.281(1)	0.468(1)	0.258(3)	H2-5-3	0.239(1)	0.470(1)	0.202(2)
C2-6	0.370(1)	0.332(2)	0.159(2)	H2-6-1	0.405(1)	0.423(2)	0.156(2)
C2-7	0.132(1)	-0.117(1)	0.245(2)	H2-6-2	0.336(1)	0.331(2)	0.093(2)
C2-8	0.144(2)	-0.204(3)	0.348(2)	H2-6-3	0.401(1)	0.234(2)	0.162(2)
C2-9	0.058(1)	-0.048(1)	0.255(2)	H2-8-1	0.107(2)	-0.293(2)	0.356(2)
C2-10	0.139(1)	-0.208(2)	0.163(2)	H2-8-2	0.133(2)	-0.127(2)	0.406(2)
H1-1-1	0.030(1)	0.439(2)	0.361(2)	H2-8-3	0.198(2)	-0.241(2)	0.354(2)
H1-2-1	0.033(1)	0.436(2)	0.144(2)	H2-9-1	0.016(1)	-0.128(1)	0.249(2)
H1-4-1	0.002(2)	0.279(4)	0.474(2)	H2-9-2	0.052(1)	0.030(1)	0.197(2)
H1-4-2	0.050(2)	0.223(4)	0.580(2)	H2-9-3	0.055(1)	0.004(1)	0.326(2)
H1-4-3	0.046(2)	0.112(4)	0.474(2)	H2-10-1	0.092(1)	-0.286(2)	0.165(2)
H1-5-1	0.097(2)	0.522(3)	0.464(2)	H2-10-2	0.190(1)	-0.262(2)	0.168(2)
H1-5-2	0.189(2)	0.473(3)	0.467(2)	H2-10-3	0.136(1)	-0.149(2)	0.094(2)
H1-5-3	0.134(2)	0.448(3)	0.573(2)				

4. Discussion

Perspective view of bis(glyoxal-bis-tert.butylimine) nickel(0) with the numbering scheme is given in Figure 1. Figure 2 shows a stereo plot of the molecule. Bond distances and bond angles are given in Table 3.

The crystal structure (Fig. 3) consists of diskrete molecules of the title complex. In the crystal exist no interatomic distance smaller as 3.5 Å.

The both glyoxal-bis-tert. butylimine ligands are connected with the nickel atom about nitrogen atoms. Two planar nickela five-membered rings are formed. Divergences from the least-squares plane are only ± 0.007 Å. The two planes included an angle from 88.95°. Central atom has nearly an exact tetragonal coordination geometry. The nickel-nitrogen distances are slightly shorter in comparison with knowned Ni—N(sp²)-bond lengths in other nickel complexes for instance Schiffsche basic (Holm, O'Conner) or nickel-2.2'-bipyridine complexes (Dinjus et al.; Sieler et al.). All nickel-nitrogen bond distances are identically within the limit of error. The short C1—1 — C1—2 and C2—1—C2—2 distances (average 1.385(20) Å) expressed a high part of double bond. Contrary to the carbon-carbon distances the four carbon-nitrogen intervals (average 1.346(15) Å) are stretched in comparison with the theoretical value of 1.30 Å (Mitchell, Cross).

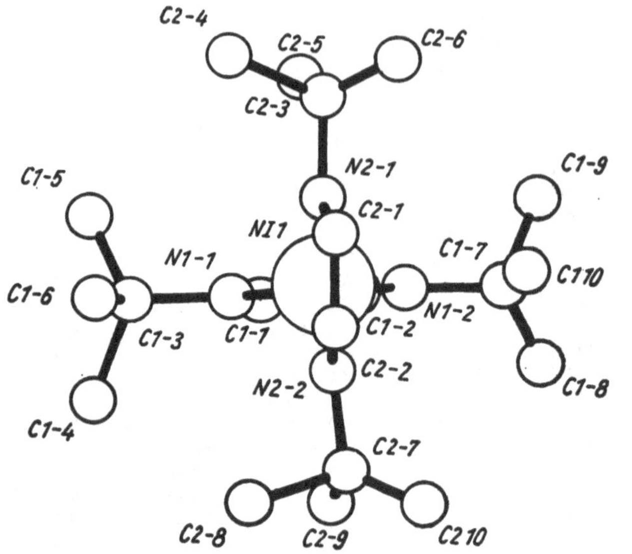

Fig. 1. The bis-(glyoxal-bis-tert.-butylimine)-nickel(0) molecule with numbering scheme

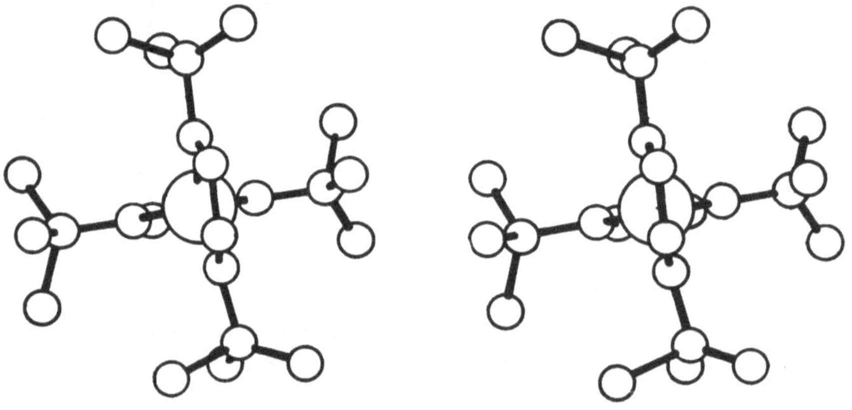

Fig. 2. Stereo PLUTO (MOTHERWELL, CLEGG) plot of the title compound

The carbon-carbon distances of the tert. butyl groups ranged from (1.41(3) to 1.63(3) Å. Avarage value is 1.53(3) Å. This is exact the theoretical worth for a $C_{sp^3}-C_{sp^3}$ single bond (NIKHAILOV). The reason for the great discrepancies are the thermal vibrations of the butyl groups.

Diamagnetism and tetragonal coordination prove this complex as a nickel(0)-complex.

The average bond distances of the three known $(dad)_2$ Ni(0) frame work are concentrated in Table 4. It means I — the title compound, II — (bis(glyoxal-bis-cyclohexylimine) nickel(0) (TOM DIECK et al. 1981), and III — bis(glyoxal-bis-2.6-dimethylphenylimine) nickel(0) (TOM DIECK et al. 1981). Almost all corresponding values are in good agreement within 3σ. The plane angles between the two dad-systems amount

Table 3

Interatomic distances (Å) and bond angles (°)

Ni1 – N1 – 1	1.91(2)	Ni1 – N2 – 1 – C2 – 3	130.8(9)
Ni1 – N1 – 2	1.93(2)	Ni1 – N2 – 2 – C2 – 2	112.5(9)
Ni1 – N2 – 1	1.918(6)	Ni1 – N2 – 2 – C2 – 7	128.8(9)
Ni1 – N2 – 2	1.925(6)	C1 – 1 – N1 – 1 – C1 – 3	116.4(21)
N1 – 1 – C1 – 1	1.36(2)	N1 – 1 – C1 – 1 – C1 – 2	110.6(29)
N1 – 1 – C1 – 3	1.51(2)	C1 – 1 – C1 – 2 – N1 – 2	120.1(28)
N1 – 2 – C1 – 2	1.33(2)	N1 – 1 – C1 – 3 – C1 – 4	107.2(21)
N1 – 2 – C1 – 7	1.49(3)	N1 – 1 – C1 – 3 – C1 – 5	108.0(22)
C1 – 1 – C1 – 2	1.41(3)	N1 – 1 – C1 – 3 – C1 – 6	118.5(24)
C1 – 3 – C1 – 4	1.63(4)	C1 – 4 – C1 – 3 – C1 – 5	118.8(29)
C1 – 3 – C1 – 5	1.48(3)	C1 – 4 – C1 – 3 – C1 – 6	100.0(24)
C1 – 3 – C1 – 6	1.53(3)	C1 – 5 – C1 – 3 – C1 – 6	104.9(26)
C1 – 7 – C1 – 8	1.44(2)	C1 – 2 – N1 – 2 – C1 – 7	117.0(23)
C1 – 7 – C1 – 9	1.58(3)	N1 – 2 – C1 – 7 – C1 – 8	112.8(29)
C1 – 7 – C1 – 10	1.55(3)	N1 – 2 – C1 – 7 – C1 – 9	111.4(23)
N2 – 1 – C2 – 1	1.36(1)	N1 – 2 – C1 – 7 – C1 – 10	98.6(23)
N2 – 1 – C2 – 3	1.51(1)	C1 – 8 – C1 – 7 – C1 – 9	105.8(30)
N2 – 2 – C2 – 2	1.33(1)	C1 – 8 – C1 – 7 – C1 – 10	120.5(28)
N2 – 2 – C2 – 7	1.49(1)	C1 – 9 – C1 – 7 – C1 – 10	107.7(28)
C2 – 1 – C2 – 2	1.36(1)	C2 – 1 – N2 – 1 – C2 – 3	116.4(11)
C2 – 3 – C2 – 4	1.57(4)	N2 – 1 – C2 – 1 – C2 – 2	114.0(11)
C2 – 3 – C2 – 5	1.52(1)	C2 – 1 – C2 – 2 – N2 – 2	117.4(12)
C2 – 3 – C2 – 6	1.46(3)	N2 – 1 – C2 – 3 – C2 – 4	96.6(19)
C2 – 7 – C2 – 8	1.62(4)	N2 – 1 – C2 – 3 – C2 – 5	103.2(12)
C2 – 7 – C2 – 9	1.52(1)	N2 – 1 – C2 – 3 – C2 – 6	112.4(24)
C2 – 7 – C2 – 10	1.41(4)	C2 – 4 – C2 – 3 – C2 – 5	109.9(14)
N1 – 1 – Ni1 – N1 – 2	83.5(3)	C2 – 4 – C2 – 3 – C2 – 6	120.4(29)
N1 – 1 – Ni1 – N2 – 1	123.9(11)	C2 – 5 – C2 – 3 – C2 – 6	101.1(25)
N2 – 1 – Ni1 – N2 – 2	83.6(4)	C2 – 2 – N2 – 2 – C2 – 7	118.5(11)
N2 – 2 – Ni1 – N1 – 2	123.3(10)	N2 – 2 – C2 – 7 – C2 – 8	98.8(21)
Ni1 – N1 – 1 – C1 – 1	117.7(17)	N2 – 2 – C2 – 7 – C2 – 9	106.0(13)
Ni1 – N1 – 1 – C1 – 3	126.0(17)	N2 – 2 – C2 – 7 – C2 – 10	116.8(23)
Ni1 – N1 – 2 – C1 – 2	110.2(17)	C2 – 8 – C2 – 7 – C2 – 9	103.4(26)
Ni1 – N1 – 2 – C1 – 7	132.8(19)	C2 – 8 – C2 – 7 – C2 – 10	113.3(15)
Ni1 – N2 – 1 C2 – 1	112.2(9)	C2 – 9 – C2 – 7 – C2 – 10	116.3(27)

to complex II 88.3° and complex III only 44.5°. That means complex III tends to a planar coordination geometry.

The crystal structures of complexes I–III show that the hypothese of the stereoscopic and electronic influences of the coordination geometry is empower.

A detail discussion about the structural data is not possible. The estimated standard deviations for complexes I and III vary from 0.005–0.03 Å for the bond lengths and from 0.1–3.0° for the bond angles.

Table 4

Average bond distances of the three known $(dad)_2Ni(0)$ frame work

compound	distance		
	Ni–N	C–C	N=C
I	1.923(11)	1.385(20)	1.346(15)
II	1.924(2)	1.400(4)	1.320(4)
III	1.928(30)	1.374(30)	1.341(30)

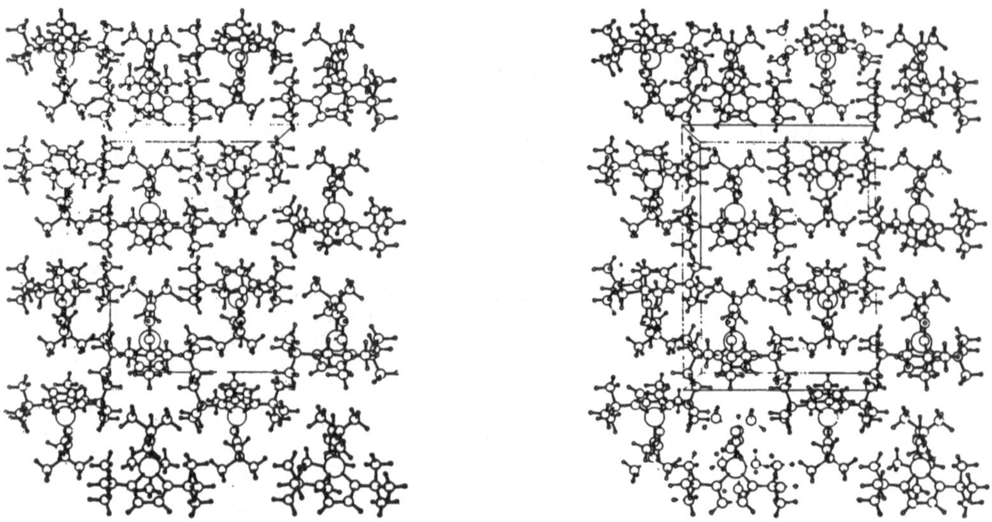

Fig. 3. Molecular packing in the crystal

The authors want to thank Mr. Dunaj-Jurco, Miss Soldanova, and Mr. Savy from the Department of Inorganic Chemistry Bratislava for the data collection, correction and the data transformation in a SHELX-76 kompatible form.

References

Balch, A. L., Holm, R. H.: Amer. Chem. Soc. **80**, 5201 (1966)

Dinjus, E., Görls, H., Uhlig, E., Walther, D., Sieler, J., Lindquist, O., Andersen, L.: J. Organomet. Chem. **276**, 257 (1981)

Holm, R. H., O'Conner, M. J.: Progr. Inorg. Chem. (St J. Lippard ed.) Vol. **14**, p. 241, New York 1971

Mitchell, A. D., Cross, L. C.: Tables of Interatomic Distances and Configuration in Molecules and Iones, Special Publ. No. 11, Chem. Soc., London 1958

Nikhailov B. M.: Invest. Akad. Nauk USSR, Otdelenie Khimitcheskij Nauk **8**, 1379 (1960)

Motherwell, W. D. S., Clegg, W.: PLUTO 78. A program for plotting molecular and crystal structures

Sheldrick, G.: SHELX 76. A program for crystal structure determination, University Chemical Laboratory, Cambridge, England 1976

Sieler, J., Dinjus, E., Walther, D., Kaiser, J., Thanh, N. N.: J. Organomet. Chem. **236**, 123 (1982)

Taper: A program for crystal structure determir ation of XTL Syntex

Tom Dieck, H., Franz, K. D., Hohmann, F.: Chem. Ber. **108**, 163 (1975)

Tom Dieck, H., Svoboda, M., Krüger, C., Tsay, Y.-H.: Z. Naturforsch. **36b**, 814 (1981)

Tom Dieck, H., Svoboda, M., Greiser, T.: Z. Naturforsch. **36b**, 823 (1981)

Tom Dieck, H., Diecks, R.: Chem. Ber. **118**, 428 (1985)

Walther, D.: Z. Anorg. Allg. Chem. **405**, 8 (1974)

Walther, D., Dinjus, U.: Z. Chem. **15**, 196 (1975)

Walther, D.: Z. Anorg. Allg. Chem. **431**, 17 (1977)

(Received February 24, 1987)

Authors' addresses:

Dr. H. Görls
Friedrich-Schiller-Universität Jena, Sektion Physik
Max-Wien Platz 1
Jena
DDR-6900

Dr. D. Walther
Friedrich-Schiller-Universität Jena, Sektion Chemie
Am Steiger 3
Jena
DDR-6900

Dr. J. Sieler
Karl-Marx-Universität Leipzig, Sektion Chemie
Liebigstr. 18
Leipzig
DDR-7010

Book Review

G. E. MURCH, M. DUPRAT (eds)

Electrochemical Methods in Corrosion Research

Band 8 in der Reihe Materials Science Forum

Trans. Tech. Publications, Aedermannsdorf (Schweiz) 1986, 620 Seiten, Preis SFr 184.00
ISBN 0-87849-550-9.

Der vorliegende Tagungsband enthält die Beiträge des zweiten Symposiums gleichen Titels, welches vom 09. bis 12. Juli 1985 in Toulouse abgehalten wurde. Diese Tagung, wie eine Vielzahl internationaler Konferenzen ähnlicher Thematik in den letzten Jahren, unterstreicht die gewachsene Bedeutung elektrochemischer Experimentaltechniken innerhalb der Korrosionswissenschaft und Korrosionsschutzpraxis. Die 55 Tagungsberichte, die sich auf Autoren aus 19 Ländern verteilen, werden unter sechs Themenschwerpunkten zusammengestellt:

1. Stofftransport und hydrodynamische Einflüsse auf die Korrosion und den Korrosionsschutz (7 Beiträge, 79 Seiten);
2. Elektrochemische Aspekte lokaler Korrosionsvorgänge (9 Beiträge, 95 Seiten);
3. Elektrochemische Methoden zum Studium der Spannungs- und Schwingungsrißkorrosion (7 Beiträge, 76 Seiten);
4. Mechanistische Aspekte des Korrosionsschutzes durch Inhibitoren und Beschichtungen (11 Beiträge, 130 Seiten);
5. Entwicklung neuer Methoden zum Studium von Korrosions- und Korrosionsschutzmechanismen (11 Beiträge, 125 Seiten);
6. Elektrochemische Methoden für verschiedenste Korrosionsuntersuchungen, insbesondere in bezug auf metallurgische Aspekte (10 Beiträge, 103 Seiten).

Der Vortragsteil wird ergänzt durch ein Sach- und ein Autorenverzeichnis (4 Seiten).

Aus dem Inhalt seien als Beispiele für die Aktualität und den trendaufzeigenden Inhalt einer ganzen Reihe von Beiträgen insbesondere solche hervorgehoben, die sich mit der elektrohydrodynamischen (Stofftransport) und mechanischen (Spannungsrißkorrosion) Impedanz und ihrer Messung, mit der Rauschanalyse (Lochfraß, Spannungsrißkorrosion, Blisterbildung), mit der Stochastik lokaler Korrosionsvorgänge und mit der Kopplung elektrochemischer mit optischen (Ellipsometrie, Vielkanalanalyse) bzw. spektroskopischen (Raman, Mößbauer) Methoden befassen.

Insgesamt reicht auch in diesem Band die Palette der Beiträge von mehr stoff- bzw. phänomen-orientierten Vorträgen bis zu vorrangig methoden-orientierten Vorträgen. Der Schwerpunkt liegt jedoch deutlich auf letzteren. Die Beiträge richten sich an Fachexperten und sind gut verständlich. Die Druckqualität ist sowohl im Text als auch bei den erfreulich zahlreichen Abbildungen ausgezeichnet. Das Buch kann jedem empfohlen werden, der an einem detaillierten Einblick in die Experimentaltechnik und Forschung auf dem Gebiet der Korrosion und des Korrosionsschutzes interessiert ist. W. FORKER

Cryst. Res. Technol.	22	1987	9	1153–1157

R. Assenov, V. Moshnikov, I. Saunin, D. Yaskov

Semiconductor Physics and Technology Research Laboratory, University of Sofia, and
The Leningrad Electrotechnical Institute, Leningrad, USSR

X-ray Study of the Reaction Products in Obtaining Lead Telluride by a Low-temperature Iodide Method

Results from X-ray studies of the products of the reaction in obtaining lead telluride by the iodide method are presented. It is demonstrated that PbI_2 condenses at the flow of the inert gas on the cool parts of the reactor at the isothermal annealing of the synthesized material. This is in agreement with the assumption about the self-alloying of PbTe as a result of the higher solubility of Pb under conditions of dissolving PbI_2 in lead telluride.

Приведены результаты рентгенодифрактометрических исследований продуктов реакции получения теллурида свинца по иодидной методике. Показано, что при изотермическом отжиге синтезируемого материала в потоке инертного газа в холодных частьях реактора конденсируется PbI_2. Это согласуется с предположением о самолегировании PbTe из-за повышенной растворимости Pb в условиях растворения PbI_2 в теллуриде свинца.

In connection with the requirements on the electrophysical characteristics of materials of optoelectronic equipment, increasing attention has been paid in recent years on the technological methods providing for a radical solution of this problem. One of these methods, in our opinion, is the low-temperature iodide method. Its application for obtaining crystals of the PbSnTe system has been carried out by a number of authors recently (Kwestroo; Gafni et al.; Moldovanova et al. 1980; Assenov et al.). The essence of this method of obtaining PbTe lies in the following:

Pb, Te, and I as elementary components are fed in a ratio ensuring the obtaining of the reaction products $PbTe:PbI_2 = 2:1$. The reaction is brought to the formation of PbTe in the PbI_2 solvent by their joint heating to 600 °C in a quartz ampoule evacuated to 1×10^{-5} Torr. The ampoule is subjected in advance to special chemical purification and to long thermal treatment at 1000 °C for the removal of residual gases (water vapours, O_2, N_2, CO, and CO_2) which, in principle, lead to deterioration of the semiconductor properties of narrow-band semiconductors (Moldovanova). Conditions are created, using a vibrator, for non-interrupted course of the reaction, taking into account the exothermal nature and the growth with the probability for explosion at the formation temperature of PbI_2 ($T = 386$ to 412 °C).

The blend obtained is crushed in agate mortar and a finely disperse powder of a brownish-gray colour is obtained. The powder is placed in a boat in a reactor purified in the same manner as the ampoule. By means of a gas flow of argon or of hydrogen specially purified through palladium diffuser, we distilled off the lead oxide over a period of 20 to 30 hours at temperatures of 500 °C to 550 °C. During the crude distillation of the iodide the boat was taken out several times from the cooled reactor and the sintered material was again crushed for the purpose of facilitating the process

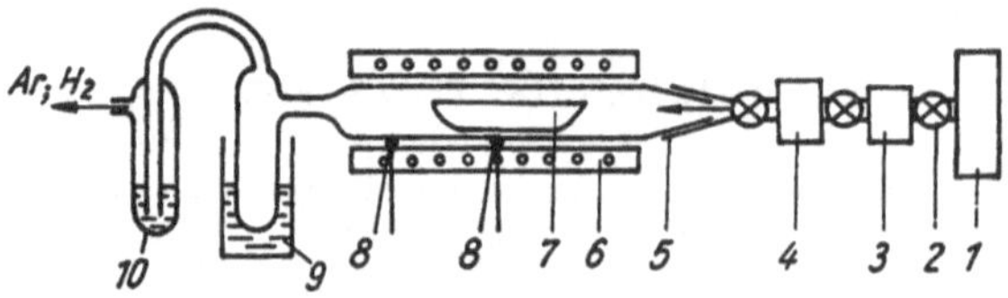

Fig. 1. Scheme of the experimental equipment for preparing of $Pb_{1-x}Sn_xTe$ by a low-temperature iodide method. 1 H_2 or Ar reservoir; 2 gas valve; 3 palladium diffuser; 4 flowmeter; 5 quartz reactor; 6 one section furnace; 7 quartz boat with the powder mixture of tin-tellurium-iodine or lead tellurium-iodine, respectively; 8 thermocouples; 9 liquid N_2 trap; 10 bubbler

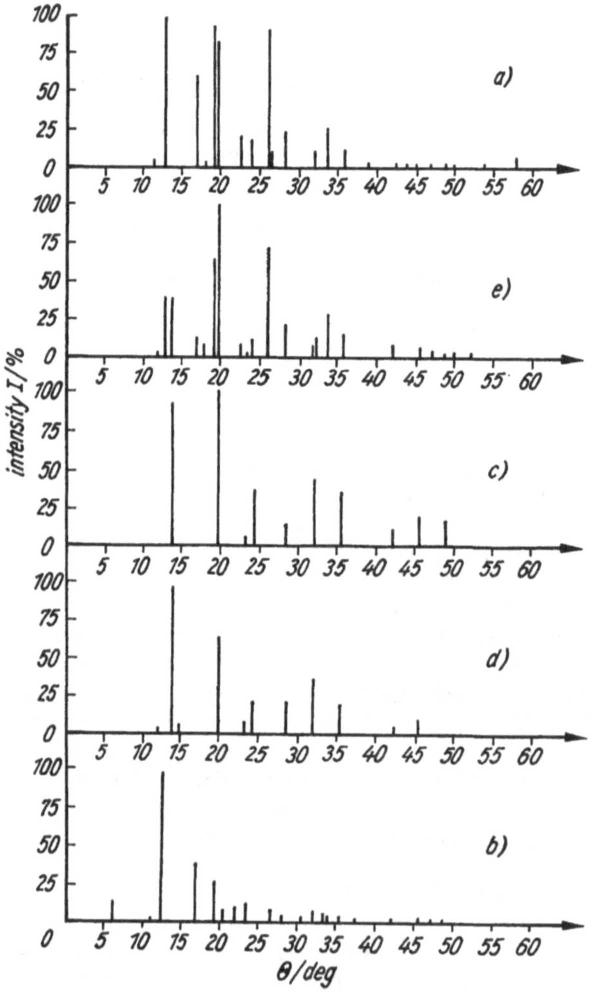

Fig. 2. Results of the X-ray phase analysis: a) Reflexes of PbI_2 crystals deposited in the reactor around the boat; b) Reflexes of PbI_2 according to tabular data (Mirkin); c) and d) Reflexes of heat-treated PbTe(I) (case c before melting ($a = 6.4671$ Å) and d) after brief melting ($a = 6.4609$ Å); e) Reflexes of PbI_2- and PbTe(I) crystals deposited at the end of the reactor

of diffusion of the iodide from the charge. The remaining solvent is controlled by weighing on analytical scales. Fine purification is done under dynamic vacuum of 1×10^{-6} Torr for 3—4 hours at 500 to 550 °C as long as the separation of the iodide continues.

The finely disperse powder obtained, which is already of a metallic-gray lustre, is placed in an evacuated ampoule and is briefly melted at 950 °C for 10—15 minutes for the formation of a monolithic ingot. The latter is sliced into thin lamellae and subjected to isothermal annealing in a flow of argon or hydrogen at a temperature of 500—550 °C for 100 hours (ASSENOV et al.).

The iodide method is of interest, compared to the classical high-temperature methods of direct alloying of the components, in view of the following advantages:

In it the reaction temperature constitutes from one-half to two-thirds of the temperature of direct synthesis, and the period of heating at this low temperature is several times lower. PbTe synthesis takes place in a liquid PbI_2 solvent and, consequently, O_2 has no direct access to PbTe. In direct PbTe synthesis we obtain the "p" type, and only after long thermal treatment the "n" type is obtained, whereas with the iodide method the crystals obtained were always of the "n" type and when they were obtained in excess of iodine they reached $n = 5 \cdot 10^{19}$ cm^{-3} and $\mu_{300K} = 900$ cm^2/V.S.

A paper of ASSENOV et al. reports on X-ray diffractometric investigations of the reaction products in obtaining SnTe by the low- temperature iodide method. It was established that not SnI_2 but SnI_4 is released in the process of temperature annealing. That is why we considered it necessary to carry out analogous investigations with PbTe obtained by the low-temperature iodide method.

The aim of our present work was to undertake direct X-ray studies of the composition and structure of the material obtained at different stages in the formation of PbTe. In our opinion it was of interest to study the composition of the substance released in the process of annealing in a flow of inert gas—argon. It appeared as flakes of a dark-yellow colour, which grew on the walls of the reactor in its coolest parts. A diagram of the reactor with the boat containing the initial substance is shown in Figure 1.

The X-ray phase analysis is carried out on an X-ray diffractometer Dron 2.0. The source of radiation used is an X-ray tube with copper anode. In order to improve sensitivity, CuK_β-radiation was not subjected to filtration. The spectra are taken from powder samples. The diffractogram shows that the phase which had grown as flakes consists of PbI_2, according to MIRKIN. X-ray studies were also made of powder samples from the blend after isothermal annealing in a flow of Ar for 30 hours (synthesis No. 16 — RD No. 3), and after brief alloying (synthesis No. 15 — RD No. 4). In either case a well-formed crystal lattice of PbTe was established. This can be ascertained by using Wulf-Bragg's relation, with $\lambda_{CuK\alpha} = 1.54178$ Å, $\lambda_{CuK\beta} = 1.39217$ Å and $a_{0\,PbTe} = 6.46$ Å : $2d \sin \Theta = n\lambda$.

On the diagrams there are no reflexes connected with the PbI_2, or with any other phase besides PbTe. This testifies to the effective separation of iodides from the initial material in the process of annealing, despite the fact that dynamic vacuum has not been used. The lattice parameter is determined from the doublet of the crystallographic plane (820) and corresponds to the value of $a_0 = 6.4671$ Å for the unmelted material in Ar flow and to $a_0 = 4.6409$ Å for the powder melted (with the aim of making a compact ingot), in Ar flow, respectively.

With the aim of increasing the probability of the appearance of other compounds which may form in the reactor but which are different from PbTe and PbI_2 in the physico-chemical conditions of condensation, we took samples both from around the boat and at the end of the reactor. A comparison between the diagrams shows that only PbI_2 is separated around the boat, whereas at the end of the reactor there are appreciable quantities of PbTe. The process of distillation of the iodide from the charge in a flow of inert gas must be continued until the first traces of PbTe (a gray film appear) at the end of the reactor, this being a visual criterion for optimization of the process of iodide distillation. It is precisely with the aim of detecting the sedimentation of PbTe and PbI_2 that a nitrogen trap has been provided at the end of the reactor (Fig. 1).

Our results in investigating the products of the reaction in obtaining SnTe (I) by the low-temperature iodide method (ASSENOV et al.) showed the existence of reflexes of the SnO_2 phase. This could take place in the process of fine purification of the iodide powder by means of dynamic vacuum, or of uncontrollable presence of O_2 in the Ar during the crude purification. That is why in the course of our present work we did not resort to dynamic vacuum in iodide distillation, and the argon used according to the certificate had O_2 not more than 5×10^{-4} v.p.c. That is why the X-ray diffractograms of the unmelted and melted PbTe (I) showed no other reflexes but those belonging to the PbTe phase. For the PbTe obtained by the direct high-temperature the lattice parameter was $a_0 = 6.460$ Å according to BIS and DIXON and $a_0 = 6.462$ Å according to DALVEN. The change in the lattice parameter in the process of heat-treatment of PbTe (I) to the value of $a_0 = 6.4671$ Å and as a result of melting in an argon flow to $a_0 = 6.4609$ Å testifies to decrease in the value of a_0 with the decrease of the iodide content in the PbTe lattice. This correlates with the data published by DEDEGKAEV et al. The effect of the lead telluride with metal component is in agreement with the conclusions drawn from the quasichemical model examined by ASSENOV et al. In this model the self-doping of PbTe (I) obtained by the iodide method is conditioned by the increase in the dissolved lead in the sublattice of the metal, resulting from the vacancies created at the dissolution of PbI_2 into PbTe.

Conclusions

1. It was established that, in the process of thermal treatment of lead telluride obtained by the iodide method, there occurs separation of iodide in the form of PbI_2. (Ih the case of tin, as earlier indicated by ASSENOV et al., this takes place in the form of SnI_4).

2. The exclusion of the stage of treatment in dynamic vacuum leads to the exclusion (or to decrease below the limit of sensitivity of the X-ray diffraction method) of the content of lead oxides.

References

ASSENOV, R., MOSHNIKOV, V. A., SAUNIN, I. V., YASKOV, D. A.: Cryst. Res. Technol. **21**, 1553 (1986)

ASSENOV, R., MOSHNIKOV, V. A., YASKOV, D. A.: in: Proceedings Mezhdunarodnaya konferentziya po elektronnoluchevim tekhnologiyam, Varna 1985, p. 447

ASSENOV, R., MOSHNIKOV, V. A., YASKOV, D. A.: phys. stat. sol. a **88**, K27 (1985)

BIS, R. F., DIXON, I. R.: J. Applied Physics **40**, 1918 (1969)

DALVEN, R.: Infrared Physics **9**, 141 (1969)

DEDEGKAEV, T. T., ZHUKOVA, T. B., LAGKUEV, D. H.: Tesisi V Vsesoyuznoi konferenzii po khimi, fiziki i tekhnicheskomu primeneniyu khalkogenidov, Baku 1979, p. 31

GAFNI, G., RAFAELOFF, R., SHAMIR, I.: Mat. Res. Bull. Vol. **14**, 217 (1979)

KWESTROO, W.: Poluchenie khalkogenidov i pniktidov pri niskoi temperature, in: Preparativnie metodi v khimii tverdovo tela, ed. P. Hagenmuller, Moscow 1976, p. 592

MIRKIN, L. I.: Spravochnik po rentgenostrukturnomu analizu polykristallov, Moscow 1961

MOLDOVANOVA, M.: in: Proceedings Khimicheskaya svyaz v kristallakh, Minsk 1969, p. 327

MOLDOVANOVA, M., TRIFÓNOVA, E. P., ASSENOV, R., KARAGIOZOV, L.: phys. stat. sol. (a) 58, K47 (1980)

(Received February 12, 1987)

Authors' addresses:

R. ASSENOV
Semiconductor Physics and Technology Research Laboratory
University of Sofia
Boulvd. A. Ivanov, No 5
Sofia 1126, Bulgaria

V. MOSHNIKOV, I. SAUNIN, D. YASKOV
The Leningrad Electrotechnical Institute, Leningrad, USSR

Book Review

LANDOLT-BÖRNSTEIN

Zahlenwerte und Funktionen aus Naturwissenschaften und Technik

Neue Serie. Gesamtherausgeber: O. MADELUNG
Gruppe III: Kristall- und Festkörperphysik
Band 7. Kristallstrukturdaten anorganischer Verbindungen. W. PIES, A. WEISS.
Teil d. Schlüsselelemente: Si, Ge, Sn, Pb; B, Al, Ga, In, Tl; Be
d 1: Schlüsselelemente Si, Ge, Sn, Pb
d 1γ: Schlüsselelemente Ge, Sn, Pb
Herausgeber: K.-H. HELLWEGE; A. M. HELLWEGE

Springer-Verlag Berlin, Heidelberg, New York, Tokyo. 1986. XXV + 215 Seiten. DM 630,–. ISBN 3-540-16169-4

Dieser Teilband schließt die Strukturdatensammlung ab, die mit Teil III/7a begonnen worden ist. Sie ist einheitlich nach Elementen geordnet, die allein oder mit anderen Elementen Anione bilden. Durch diese Ordnung sind chemische Verwandtschaften einfacher darstellbar. Nach den einfachen Oxoverbindungen folgen bei jedem Schlüsselelement Oxoverbindungen mit weiteren Anionen.

Folgende Kette von Verbindungen soll das Ordnungsschema am Beispiel der einfachen Oxoverbindungen des Germaniums (ohne H_2O) illustrieren: Li_4GeO_4, ... , $Li_6Ge_2O_7$, ... , Na_4GeO_4, ... , $LiNaGe_4O_9$, ... , $Cs_6Ge_2O_7$, ... , $Sm_2Ge_2O_7$, ... , $NaSmGeO_4$.

Die Verbindungen sind durchnumeriert und damit über das noch zu erwartende Substanzregister schnell auffindbar. Die wichtigsten Daten der Sammlung sind Raumgruppe, Gitterkonstanten (in Å!) mit Fehlerangaben, Massendichte, Strukturtyp, Kristallform, Angaben zur Strukturbestimmung und ausgewählte physikalische Eigenschaften. Die Literatur ist bis 1978 berücksichtigt, wobei die Quellen des Zeitraumes 1971–78 als Fußnoten in vorliegendem Band ausgewiesen sind.

Die Autoren haben sich der Mühe unterzogen, Bestwerte unter den Quellenangaben auszuwählen, das umfangreiche Material also wertend zu sichten.

Mit rund 25000 so erfaßten Verbindungen in den Teilbänden a–f erwerben sich die Autoren ungewöhnliche Verdienste um die quantitative Festkörperforschung, die von zuverlässigen Strukturdaten lebt. Auch wenn künftig elektronische Datenbanken den Umgang mit dieser Zahlenfülle erleichtern werden, den Vorzug des bequemen Zugriffs in einer gut sortierten Fachbibliothek behalten die Bände allemal.

Und überdies profitieren auch Datenbanken von der aufgewandten Mühe. Den Autoren und dem Verlag gebührt hohe Anerkennung für diese Leistung. P. PAUFLER

Cryst. Res. Technol.	22	1987	9	1159–1163

R. ASSENOV, V. A. MOSHNIKOV, D. A. YASKOV,
A. P. ANDREEV, M. A. BESTAEV

Semiconductor Physics and Technology Research Laboratory, University of Sofia;
The Leningrad Electroctechnical Institute, USSR

Diffusion of Iodine in Lead Telluride

By using the non-stationary thermal-probe method, investigations were carrieed out on the distribution of the coefficient of thermal electromotive force along the depth of monocrystals of lead telluride, subjected to isothermal saturating annealing in vapours of alloying compositions. The alloying composition consisted of powdered lead telluride, synthesized by the iodide method, and contained elevated concentration of the metallic component and iodine. The PbTe monocrystals were obtained from the vapour phase and had a natural cut. The results of the investigations testify to considerable changes in the concentration of the electrically active proper defects within the entire volume of the crystals, this being obviously connected with diffusion of vacancies from the crystal volume.

Использованием нестационарного термозондового метода проведены исследования распределения коэффициента темро-э.д.с. по глубине монокристаллов теллурида свинца, подвергнутых изотермическому насыщающему отжигу в парах лигатуры. Лигатура состояла из порошка теллурида свинца, синтензированного по иодидной методике, и содержала повышенную концентрацию металлического компонента и иода. Монокристаллы PbTe были получены из паравой фазы и имели естественную огранку. Результаты исследований сведетельствуют о значительном изменении концентрации электрически активных собственных деффектов по всему объему кристаллов, что, по-видимому, связано с диффузией вакансий из объема кристаллов.

Many investigations (FRITTS; BOIKOV et al.; DEDEGKAEV et al.) connected with obtaining materials of n-type conductivity for thermoelectrical purposes have been dedicated to the application of iodine as alloying admixture in lead telluride.

Interest in iodine behaviour in PbTe has again become intensified in recent years in connection with the development of the methods of obtaining chalcogenides in iodide solvents (KWESTROO; MOLDOVANOVA et al.; GAFNI et al.; ASSENOV et al.). Possessing a number of valuable technological characteristics connected with reducing the temperature and the purification of foreign admixtures (MOLDOVANOVA et al.), the iodide method also introduces essential changes in the electrophysical properties. Publications by ASSENOV et al. examine various models explaining the observed change in the electrophysical properties.

In this connection it was of interest to study the changes in the electrophysical properties in the process of treatment in vapours of alloying admixtures.

The aim of the work undertaken was to investigate the distribution of the coefficient of the thermal electromotive force α in the depth of PbTe samples subjected to isothermal annealing in vapours of the PbTe charge, obtained by the iodide method.

1. Method of obtaining and of treatment of PbTe samples

PbTe monocrystals were grown from a vapour phase in closed quartz ampoules in two-section tubluar furnaces. The crystals had the natural cut along plane (100). The value of the Hall concentration of charge carriers corresponded to $P_{300K} = 5 \times 10^{18} \dots 10^{19}$ cm^{-3}. The monocrystal was evaluated by X-ray topography. The dislocation density was determined by etch pits and was equal to $\sim 10^4$ cm^{-2}.

The technological way of obtaining the PbTe charge by the iodide method involved the following:

Pb, Te, and I are charged as elementary components at a ratio ensuring the obtaining of the reaction products PbTe:PbI_2 = 2:1. The reaction is brought to the formation of PbTe in the PbI_2 solvent by joint heating at about 600 °C for 20—30 hours in a quartz ampoule evacuated up to 1×10^{-5} Torr, which has been chemically and thermally treated in advance. Using a vibrator, conditions were created for continuous course of the reaction and, taking account of its exothermal character, retentions in the rise of the temperature, suitable in their duration and places, were carried out.

The charge obtained was crushed in an agate mortar and a finely dispersed powder of brownish-gray colour was obtained. PbI_2 was separated in a reactor by using a flow of pure Ar at 550 °C. During this crude separation of PbI_2 the boat with the powder was taken out several times from the cooled reactor and the sintered material was crushed anew in order to facilitate the process of diffusion of iodide from the charge.

In view of the conclusions contained in a publication of Assenov et al. and of the aim of the present work, no fine purification of the charge from the iodide in dynamic vacuum was carried out. The possibility is created in this manner for performing diffusion at appropriate temperature and time.

The single-phase character of the charge was checked using the X-ray diffractometer Dron-2.0. The analyzing radiation used was CuKα. The method and the results of the analysis of the reaction products have been described in detail by Assenov et al. The charge used was powdered PbTe $\langle I \rangle$ possessing n-type conductivity and, consequently, containing a higher quantity of lead than the stoichiometric one. The electrophysical properties were controlled by the thermal probe. The excess of metal in PbTe corresponded to the traditional conditions of isothermal annealing of samples by Brebrick's method (Melngailis, Harman) which makes it possible to remove the surface layers at the boundary of the homogeneity area from the part of the metal at the thermal treatment. A difference of principle was the presence of iodine in the alloying composition.

The phase contact was realized only through the gas phase. Double quartz ampoules were used to that aim. The smaller ampoule contained the samples investigated, while the outer ampoule contained the charge whose mass was 5—6 times larger than the crystal processed. Prior to the isothermal annealing the ampoules were evacuated to a residual pressure of $\sim 10^{-5}$ Torr. Annealing was carried out in resistor furnaces (type SUOL). The temperature was regulated by means of diagrams based on the high-accuracy temperature regulator BPT-3. The temperature of the isothermal annealing in the experiments varied from 500 °C to 750 °C. Annealing time in all experiments amounted to 10 hours.

2. Method of measurement

The thermal-probe method was used for investigating the electrophysical properties in the diffusion process. The application of this method for the analysis of PbTe samples is determined first of all by the high values of the coefficient of thermal electromotive force and by the low values of the electrical resistance of the samples investigated and also by the possibility of obtaining information in the local areas of the sample.

By way of shortcomings of the method of thermal electromotive force it is necessary to refer to the complexity of separating the influence of the inhomogeneity of the composition (or the concentration of the electrically active defects) and the influence of the fluctuations in the temperature below the tip of the probe for the values of the coefficient of the thermal electromotive force α_T. Furthermore, the calculation of α_T as increase in the value of the thermal electromotive force ΔU_T in relation to the temperature difference $\Delta T(\alpha_T = \Delta U_T/\Delta T)$ is correct only if α_T does not depend on the temperature. These conditions are not satisfied in the area of mixed conductivity.

In order to exclude the above errors we adopted the non-stationary thermal-probe method (MOSHNIKOV; ORMONT). Essentially, the method consists of taking the dependence $U_T = f(\Delta T)$ and of determining the coefficient of thermal electromotive force α_T by graphic differentiation. In this case the value of α_T is found for the concrete value of the temperature, regardless of the character of the curve $U_T = f(\Delta T)$. The diagram of the equipment used for measuring the α_T is shown in Figure 1. The equipment consists of probe-thermocouple, a block for maintaining the temperature, and a massive brass electrode mounted on the bed-plate of the optical microscope. A zero-thermostat is used for stabilization of the outer junctions of the thermocouples. The optical part of the microscope is used for visual selection of the place of contact between the probe-thermocouple and the surface of the sample. The temperature of the heater was smoothly regulated by changing the voltage with an in-series-connected step-down transformer and laboratory-type autotransformer. Calculating the coefficient of thermal electromotive force α_T under non-stationary conditions was done by means of the curve $\Delta U_T = f(\Delta T)$, automatically registered on a two-coordinate plotting board by a recording potentiometer PDP-4.

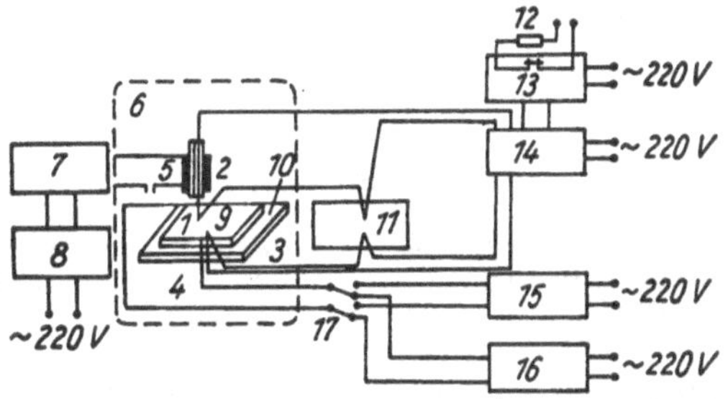

Fig. 1. Diagram of the equipment for measuring the coefficient of thermal electromotive force α_T by the non-stationary thermal-probe method. 1, 2 measurement thermocouple, 3, 4 control thermocouple, 5 heater, 6 box of acrylic plastic, 7, 8 step-down transformers, 9 sample, 10 cold electrode-base plate, 11 zero-thermostat, 12 – 14 block for temperature stabilization, 15 two-coordinate recorder PDP-4, 16 recorder H-373, 17 commutator

3. Experimental results

Measurements were carried out in a direction perpendicular to the plane (100). The typical dimensions of the PbTe samples were $4 \times 3 \times 2$ mm^3. Five to six local areas were selected on the surface, in which measurements were carried out of the coefficient of thermal electromotive force α_T and of the thickness of the sample by means of a point-type thickness gauge. This was followed by taking off the surface layer, and then the measurements of the thickness and of the value of α_T were repeated. The dependence of the averaged values of the coefficient of thermal electromotive force α_T on the depth of the samples is shown in Figure 2.

It follows from the dependence $\alpha_T = f(z)$ that, upon treatment in the vapours of the alloying composition, the values obtained by the iodide method of the coefficient of thermal electromotive force α_T in the surface layers change their signs.

Under these conditions at $T = 500$ °C and $T = 600$ °C the thermoelectric properties change in the surface layers only, whereas the value of the coefficient of thermal electromotive force α_T in the depth of the sample remains equal to $+100$ µV/K. This value of α_T corresponds to the value of the coefficient of thermal electromotive force

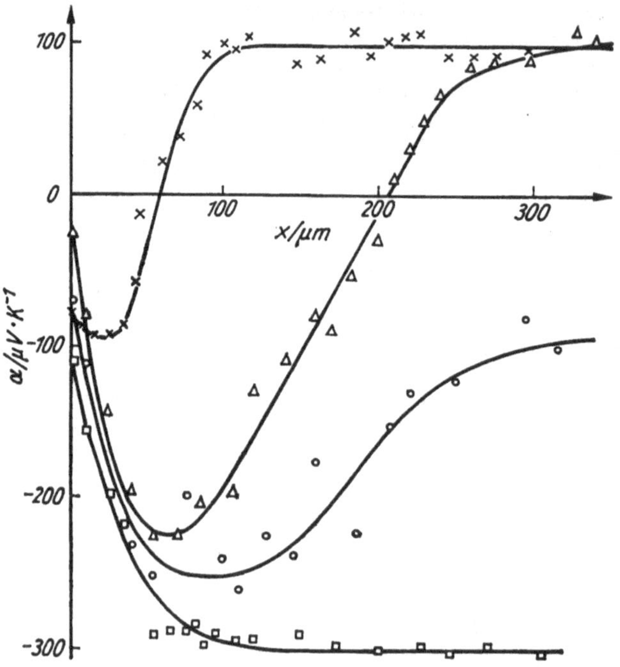

Fig. 2. Dependences of the coefficient of thermal electromotive force in depth of the PbTe samples subjected to isothermal saturating annealing at different temperatures: 500 °C (×), 600 °C (△), 700 °C (○) and 750 °C (□) (on the abscissa read z (μm)

for samples not subjected to saturating alloying in vapours of alloying compositions enriched in iodine and lead.

The results of the treatment of PbTe monocrystals at annealing temperatures of 700 °C and 750 °C reveal essential qualitative differences. Under these conditions the thermoelectric properties of the material undergo changes not only in the surface layers but also in the volumes situated at a considerable depth. At the thickness values of the monocrystals investigated by us ($\sim$ 1 mm) the changes observed take place within the entire volume of the PbTe crystals. At T = 750 °C the α_T value reaches -300 μV/K and actually does not change in the entire volume, with the exception of the surface layers. At the treatment of PbTe monocrystals there occurs a disruption of the monocrystallinity of the surface layers to a depth of $\sim 50-70$ μm. The depth of the disrupted layer grows with the increase in the value of the temperature of the isothermal annealing. As can be seen from Figure 2, this process is reflected also on the dependence $\alpha_T = f(z)$.

The experimental dependences obtained $\alpha_T = f(z)$ are in qualitative agreement with the model of structure division (ASSENOV et al.; MOSHNIKOV et al.; ORMONT) at the formation of solid solution PbTe–PbI_2. The thermodynamic description of the possibility of such a reaction taking place has been examined by PIKALEV et al. during mass-spectrometric investigations of the composition of the vapour phase above PbTe alloyed with sodium and iodine and crystallization of PbTe films from a molecular beam in vacuum. This must lead, on the one hand, to increase in the actual solubility of Pb in the PbTe structure due to the formation of excess vacancies in the metal sublattice, while on the other hand it ensures the opposite flow of Pb vacancies from the volume of the PbTe crystal toward the surface of the sample (since the equilibrium concentration of the metal vacancies is considerably higher in the layers with higher PbI_2 content). In our opinion it is precisely with this flow of vacancies that it is possible to link the change in the thermoelectric properties within the entire volume of PbTe monocrystals.

In this manner, isothermal annealings have been performed with PbTe monocrystals in vapours of charges of lead telluride enriched with iodine and lead, and an investigation was made of the change in the thermoelectric properties in the depth of the sample. An essential change was established in the value of the coefficient of thermal electromotive force α_T in the entire volume of the PbTe crystals. The dependences observed are in qualitative agreement with the model of formation of the structure of division $PbTe\langle PbI_2\rangle$ and the possible redistribution, arising from this model, of the concentration of the proper electrically active defects in the volume of the crystal.

References

Assenov, R., Moshnikov, V. A., Yaskov, D. A.: in: Proceedings Mezhdunarodnaya Konferenziya po lektronno-luchevim tekhnologiyam, Varna 1985, p. 447

Assenov, R., Moshnikov, V. A., Yaskov, D. A.: phys. stat. sol. a 88, K27 (1985)

Assenov, R., Moshnikov, V. A., Saunin, I. V., Yaskov, D. A.: Cryst. Res. Technol. 22, 1153 (1987)

Boikov, J. A., Golzman, B. M., Gudkin, T. S., Dedegkaev, T. T., Zhukova, T. B., Iordanishvili, E. K., Shmuratov, E. A.: Fiz. Tverd. Tela 22, 1226 (1980)

Dedegkaev, T. T., Lagkuev, D. H., Moshnikov, V. A., Yaskov, D. A.: Izvestya LETI, vip. 250, 103 (1979)

Fritts, R. V.: Termoelektricheskie materiali i preobrazovateli, Moscow 1964, p. 163

Gafni, G., Rafaeloff, R., Shamir, I.: Mat. Res. Bull. 14, 217 (1979)

Kwestroo, W.: Poluchenie khalkogenidov i pniktidov pri nikoi temperatur, in: Proceedings Preparativnie metodi v khimii tverdogo tela, R. Hagenmuller (ed.), Moscow 1976, p. 592

Melngailis, I., Harman, T. C.: Single Crystal Lead-Tin Chalkogenides, in: Semiconductors and Semimetals, Vol. 5, eds R. K. Willardson, A. C. Beer, New York, London 1970, p. 111

Moldovanova, M., Trifonova, E. P., Assenov, R., Karagiozov, L.: phys. stat. sol. a 58, K47 (1980)

Moshnikov, V. A., Miloslavov, S. L.: Tesisi dokl. I Vs. konferenzii Poluchenie i svoistva soedineniy tipa $A^{II}B^{VI}$ i tverdikh rastvorov na ikh osnove, 42, Moscow 1977, p. 310–311

Moshnikov, V. A.: Ph. D. Thesis, Leningradskiy Elektrotekhnicheskiy institut 1981

Ormont, B. F.: Introduction to Physical Chemistry and Crystal Chemistry of Semiconductors, Moscow 1973, p. 656

Pikalev, A. P., Merzlyakov, A. V., Ozhegov, P. I., Polistanskij, Yu. G.: in: Proceedings Termoelektricheskie materiali i plenki, Moscow 1976, p. 119–125

(Received February 12, 1987)

Authors' addresses:
R. Assenov
Semiconductor Physics and Technology Research Laboratory
University of Sofia
Boulvd. A. Ivanov, No 5
Sofia 1126, Bulgaria

V. A. Moshnikov, D. A. Yaskov, A. P. Andreev, M. A. Bestaev
The Leningrad Electrotechnical Institute
Leningrad, USSR

Book Review

I. Hargitttai (ed.)

Symmetry. Unifying Human Understanding.

Internat. Series in Modern Applied Mathematics and Computer Science, Vol. 10.

Pergamon Press, New York-Oxford-Toronto-Sydney-Frankfurt 1986. 1045 p. US $ 115.–,
ISBN 0-08-033986-7

The present book is the outcome of the fascination which symmetry has on scientists of quite different disciplines. It covers a tremendous field ranging from crystallography and physics through biology and biochemistry to music, literature and arts, and is aimed at contributing to the bridging role of symmetry in those diverse fields.

Symmetry is treated as a fundamental concept in science rather than an aesthetic phenomenon. 65 contributions are loosely arranged without grouping by specialization so that readers are stimulated to have a look upon papers remote from their own primary field of interest. Written by specialists from 18 nations the individual contributions are more monographs than chapters of a textbook and represent different approaches to the subject. Part of them give a fairly complete overview of a special area, others express more a personal point of view or belief of the author. Moreover, the issue maintains a good balance between theoretical and applied aspects. Among the attractive features are a comprehensible presentation, emphasis on fundamentals, inclusion of good examples, attractive pictures and numerous references and a high standard of printing.

Although it is unlikely that all contributions of this book will simultaneously be of use to the reader, there may be many useful associations even from more remote subjects which will bear fruit after a certain 'incubation' period. Since all papers were prepared in such a way as to appeal a wide audience, the book can be recommended to all symmetry-minded postgraduates. It represents the most extensive compilation of up-to date contributions to that subject now available. Thanks are due to the authors, the editor and the publishers for their excellent work.

P. Paufler

Cryst. Res. Technol.	22	1987	9	1165–1172

K. G. SUBHADRA, K. A. HUSSAIN

Department of Physics, Kakatiya University, Warangal, India

X-ray Determination of Thermal Expansion of $Na(ClO_3)_x(BrO_3)_{1-x}$ Mixed Crystals

Lattice constants of sodium chlorate and sodium bromate and their mixed crystals were determined over the temperature range from room temperature to 320 °C. The observed lattice constants for the various compositions show strong deviations from those expected from Vegard's rule. The lattice constants at various temperatures were fitted to polynomials of type $a = a_0 + a_1t + a_2t^2$. The linear coefficients of expansion for the $Na(ClO_3)_x \cdot$ $\cdot (BrO_3)_{1-x}$ series varies non-linearly with composition showing negative deviations from linearity.

1. Introduction

Sodium chlorate ($NaClO_3$) and sodium bromate ($NaBrO_3$) are isomorphous cubic crystals with space group $P\,2_13$. Considerable information is available regarding several of the physical properties like elastic constants (RADHA, GOPAL; SRINIVASAN, GOPAL; GLUYAS et al.), thermal expansion (SHARMA; GANESAN; DESHPANDE, MUDHOLKER; WATHORE, KULKARNI), dielectric constants (MASON; SIRDESHMUKH, REDDY) etc. Because of their particular symmetry, they are among the few cubic crystals showing optical activity (CHANDRASEKHAR, MADHAVA) and piezo-electricity (MASON). These crystals are completely miscible. But the information about the physical properties of the mixed crystals is limited. The lattice constants (KLEBER, KOLTERMANN; MOHANLAL) and the optical rotation (CHANDRASEKARAN, MOHANLAL) are, by far, the only two properties which have been studied in some detail with regard to their composition dependence.

In this paper we report the results of a systematic determination of the temperature variation of lattice constants and thermal expansion of the mixed crystals of $NaClO_3$ and $NaBrO_3$. The motivation for this study was two-fold. Firstly, there is very meagre work on the composition dependence of the thermal expansion of compound mixed crystals. Secondly, in a family of isomorphous crystals, the thermal expansion coefficient correlates with the lattice constant; the crystals with a higher lattice constant have a larger expansion coefficient. For example, the lattice constants of KCl and KBr are 6.279 Å and 6.583 Å respectively and their coefficients of expansion are 38.0×10^{-6}/°C and 40.3×10^{-6}/°C (KANTOLA). The lattice constants of $NaClO_3$ and $NaBrO_3$ are 6.5770 Å and 6.7061 Å. Their expansion coefficients at room temperature are 41.1×10^{-6}/°C and $31.5 + 10^{-6}$/°C (DESHPANDE, MUDHOLKER; WATHORE, KULKARNI). Thus, the correlation between alttice constant and expansion coefficient breaks down for this pair of crystals. Because of this situation, a study of the composition dependence of the thermal expansion of the mixed crystals of $NaClO_3$ and $NaBrO_3$ appeared to be all the more interesting.

2. Experimental

The starting materials were 99.95% pure sodium chlorate supplied by E. Merck and 99.96% pure sodium bromate supplied by Riedel-De Haan. The crystals of $NaClO_3$, $NaBrO_3$ and their mixed crystals were grown from aqueous solutions by slow evaporation. The details of crystal growth have been described in our earlier paper (SUBHADRA, HUSSAIN). In each case, a crystal was crushed into fine powder for X-ray studies and composition determination. The composition was determined by polarographic analysis. The error in the composition was determined by taking two polarograms for a given sample and was estimated to be about two per cent. Apart from the pure components, mixed crystals with six compositions were prepared and studied in this work.

X-ray powder photographs were obtained with filtered CuKα radiation employing a camera designed by SIRDESHMUKH and DESHPANDE. It is essentially a symmetric focusing camera with a diameter of 15 cm. The sample in the form of powder is filled in a cavity in the sample holder and has a flat surface. The ring diameters were measured using a comparator with an accuracy of 0.005 cm. For all the films, the ring diameters were measured at least six times to minimize random errors. From five to seven well-resolved high angle doublets for values of the Bragg angle $\Theta > 70°$ were used in the evaluation of the lattice constant.

The systematic errors arising due to displacement of the sample from the focusing circle, vertical divergence of the X-ray beam, film shrinkage and radius errors are accounted for by the function $\Phi \tan \Phi$ (COHEN) where Φ is the complement of the Bragg angle Θ. It was shown by SIRDESHMUKH and DESHPANDE that this error function also accounts for the systematic errors caused by using a flat sample instead of a curved sample. For the determination of accurate lattice constants in the present work, the lattice constants were plotted against $\Phi \tan \Phi$ and fitted to a least square line. The true lattice constant was obtained to a least square line. The true lattice constant was obtained by an extrapolation of this line to $\Theta = 90°$. The error in the lattice constant was determined by the linear regression method and was usually of the order of 0.0002 Å and was slightly higher (0.0005 Å) for mixed crystals near about the equimolal composition.

For study at elevated temperatures, X-ray photographs were recorded at six or more different temperatures in the temperature range 30 °C to 320 °C, the highest temperature studied depended on the quality of the pattern obtained and the temperatures upto which the samples were stable. The heating arrangement consists of a cylindrical heater coaxial with the sample holder. The temperature along the axis of the heater was constant except near the ends. This ensured uniform heating of the sample placed within the heater. A chromel-alumel thermocouple welded into the sample holder was used to determine the temperature of the sample. The temperatures of the samples were determined with an accuracy of ± 0.5 °C.

The data on lattice constants at different temperatures was fitted to an equation of the type

$$a_t = a_0 + a_1 t + a_2 t^2. \tag{1}$$

where t is the temperature in °C and a_t is the lattice constant at temperature t.

3. Results and discussion

The lattice constants for $NaClO_3$–$NaBrO_3$ mixed crystal system have been determined earlier by KLEBER and KOLTERMANN and by MOHANLAL. The general relationship between lattice constant and composition can be represented by the equation

$$a^n = x a_1^n + (1 - x) a_2^n , \tag{2}$$

where a, a_1 and a_2 are the lattice constants of the mixed crystal and the end members, and x is the molar fraction of the component with lattice constant a_1. Various authors have used the values 1, 3 and 8 for n. Eq. (2) with $n = 1$ is known as Vegard's rule.

SIRDESHMUKH and SRINIVAS point out that for most of the alkali halide mixed crystals, the lattice constants vary linearly with composition, although in a few cases, $n = 3$, gave a better fit. For the $NaClO_3$–$NaBrO_3$ system, whereas KLEBER and KOLTERMANN observed a linear variation, MOHANLAL found that his data do not fit Eq. (2) with any of the values of n mentioned above. As the values of lattice constants in the present investigation are more accurate than the earlier values, the relevance of Eq. (2) to this system is reexamined.

The values of the lattice constants at 30 °C obtained in the present work for $NaClO_3$, $NaBrO_3$ and their mixed crystals are given in Table 1 along with the results obtained by KLEBER and KOLTERMAN and by MOHANLAL). The value of the lattice constant for $NaClO_3$ agrees with that obtained by MOHANLAL. However, the lattice constant

Table 1
Room temperature lattice constants for $Na(ClO_3)_x(BrO_3)_{1-x}$ mixed crystal system

Sl. No.	present work		MOHANLAL		KLEBER, KOLTERMANN	
	x	a/Å	x	a/Å	x	a/Å
1	0.00	6.7074 ± 0.0002	0.00	6.716 ± 0.003	0.00	6.693
2	0.03	6.7054 ± 0.0002	0.06	6.725 ± 0.008	0.322	6.652
3	0.10	6.7008 ± 0.0002	0.07	6.727 ± 0.015	0.488	6.629
4	0.15	6.6930 ± 0.0001	0.10	6.709 ± 0.001	0.675	6.607
5	0.59	6.6396 ± 0.0003	0.22	6.708 ± 0.006	0.84	6.585
6	0.81	6.6041 ± 0.0003	0.24	6.714 ± 0.002	1.00	6.564
7	0.84	6.6032 ± 0.0003	0.24	6.703 ± 0.003		
8	1.00	6.5778 ± 0.0001	0.53	6.648 ± 0.004		
9			0.71	6.628 ± 0.005		
10			0.86	6.624 ± 0.012		
11			0.92	6.606 ± 0.003		
12			1.00	6.579 ± 0.003		

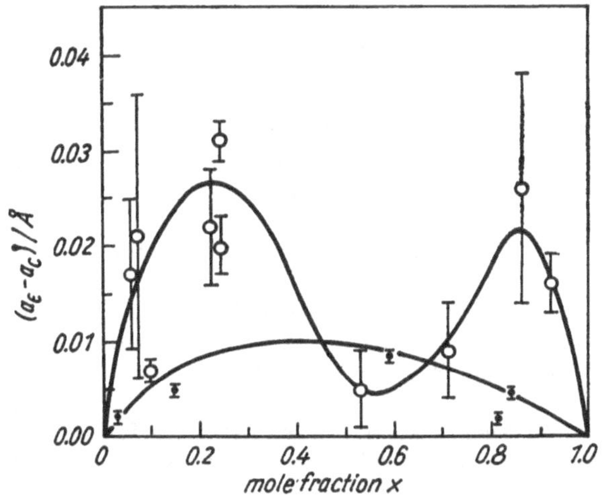

Fig. 1. Plot of $(a_E - a_C)$ against composition x for the $Na(ClO_3)_x(BrO_3)_{1-x}$ system. ○ Mohanlal, ● present work

for $NaBrO_3$ (6.7074 ± 0.0002 Å) in this measurement is lower than that obtained by MOHANLAL, but is in agreement with the value of 6.7061 Å from the more recent work of WATHORE and KULKARNI. Plots of lattice constant vs x do not clearly reveal the nature of deviations from linearity. Instead, SLAGLE and MCKINSTRY plotted differences ($a_E - a_C$) against x where a_E is the experimental value of the lattice constant and a_C that calculated from Eq. (2). In Figure 1 plots of ($a_E - a_C$) against x are given with $n = 1$ for the present results and MOHANLAL's result. Both sets reveal positive deviations from Vegard's law. SLAGLE and MCKINSTRY (1966) observed a maximum deviation of the order of 10^{-3} Å from Vegard's rule in the KCl_xBr_{1-x} series. In the present case the deviations are much larger and are of the order of 10^{-2} Å. MOHANLAL's data points lie on a double peaked curve. The data points from the present measurement lie on a single peaked curve and the values of ($a_E - a_C$) are of smaller magnitude than those observed by MOHANLAL. As in MOHANLAL's work, the present results also cannot be represented by choosing values of 3 or 8 for n in Eq. (2). In fact, as an exercise, n was allowed to take still higher values like 15, 25, and 30, and the lattice constants calculated. The best fit between experimental and calculated lattice constants is obtained for $n = 25$. Even for this value of n, the estimated standard deviation in ($a_E - a_C$) is 0.0022 which is still larger than the experimental errors in the lattice constant.

The lattice constants of $NaClO_3$ and $NaBrO_3$ at elevated temperatures are plotted in Figures 2 and 3 along with the results obtained by earlier workers. It may be seen that the present values of lattice constants of $NaClO_3$ at elevated temperatures are consistent with those of DESHPANDE and MUDHOLKER at corresponding temperatures and both sets of data fall on a curve which is a smooth extension of the low temperature data by NASR-EDDINE et al.). For $NaBrO_3$, also, the high temperature lattice constants obtained in this work and by WATHORE and KULKARNI fall on the same curve, and are a smooth extension of the low temperature values by NASR-EDDINE et al.

The values of the coefficient in the polynomial Eq. (1), for the mixed crystals with various compositions are given in Table 2. By differentiation of Eq. (1) and using the definition

$$\alpha = \frac{1}{a_{30}} \frac{da}{dt} \tag{3}$$

the coefficient of expansion can be obtained. These polynomials yield a linearly temperature dependent coefficient of expansion. The coefficients of thermal expansion of $NaClO_3$ and $NaBrO_3$ obtained in the present work and by earlier workers are given in Tables 3 and 4, respectively. The present values of thermal expansion are in better agreement with earlier X-ray values rather than from interferometric measurements.

There is very little systematic work on the composition dependence of thermal expansion of mixed crystals. The limited work on KCl–KBr (KANTOLA; SALIMAKI; POZDEEVA et al.; SLAGLE, MCKINSTRY) indicates that the composition dependence is slightly non-linear with positive deviations from linearity. This trend is qualitatively in agreement with a theory developed by HIETALA based on thermodynamic considerations. For the KCl–RbCl system IVANKINA and POZDEEVA observed a non-linear variation of thermal expansion with composition with a maximum at the equimolal composition.

The coefficients of linear expansion at room temperature for the mixed crystals in the $Na(ClO_3)_x(BrO_3)_{1-x}$ series are plotted against composition in Figure 4. It can be seen that the plot is non-linear with negative deviations from linearity. The maximum deviation from linearity is of the order of ten per cent. Thus, the composition dependence of thermal expansion in this series is different from that in the KCl–KBr

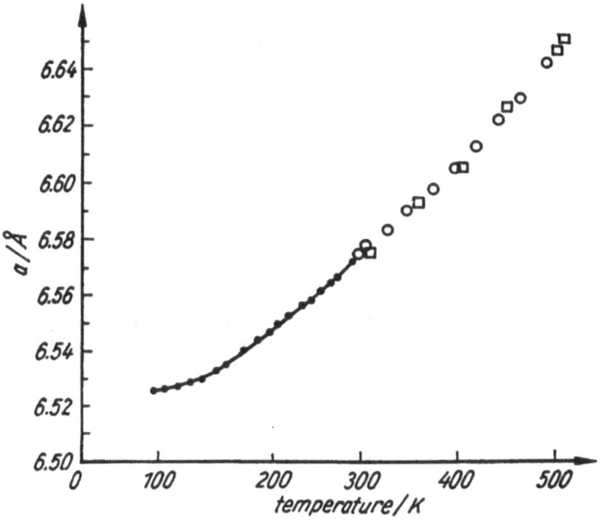

Fig. 2. Plot of lattice constant a against temperature for $NaClO_3$. ● ● ● ● NASR-EDDINE et al.; □ DESHPANDE, MUDHOLKER; ○ present paper

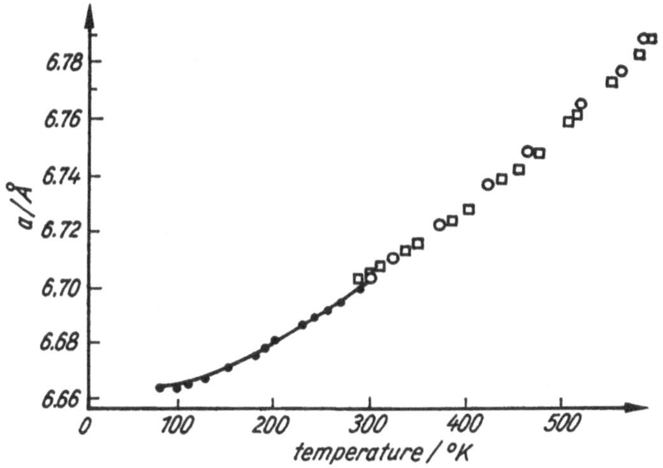

Fig. 3. Plot of lattice constant a against temperature for $NaBrO_3$. ● ● ● ● NASR-EDDINE et al.; □ WATHORE, KULKARNI; ○ present paper

series. We must hasten to mention that these two crystal pairs differ in some other respects too. Let us consider the correlation between the lattice constants and thermal expansion. As mentioned in section (1), the correlation between lattice constant and thermal expansion is entirely different in these two pairs of crystals. It has been pointed out by RADHA and GOPAL that in sodium chlorate and sodium bromate, the strength dependent properties like elastic constants and thermal expansion depend on the relative values of the Na-Cl and Na-Br distances rather than on the lattice constants. From the X-ray structural data given by ABRAHAMS and BERNSTEIN, these distances have values 3.98 Å in $NaClO_3$ and 3.82 Å in $NaBrO_3$. The Na–Br distance in $NaBrO_3$ is shorter than the Na–Cl distance in $NaClO_3$, although the lattice

Table 2

Values of the constants of the polynomial $a_t = a_0 + a_1 t + a_2 t^2$ for $Na(ClO_3)_x\,(BrO_3)_{1-x}$ mixed crystal system

x	a_0	$a_1 \times 10^4$	$a_1 \times 10^7$	temperature range studied (°C)
0.00	6.7007	2.173	2.11	26.0–306.0
0.03	6.6989	2.110	2.53	29.0–285.5
0.10	6.6945	2.021	3.19	28.0–240.5
0.15	6.6864	2.127	2.96	34.0–260.0
0.59	6.6334	1.912	5.30	32.0–149.5
0.81	6.5976	1.988	5.14	26.5–167.5
0.84	6.5963	2.119	5.33	35.5–239.5
1.00	6.5705	2.295	4.56	27.0–219.0

Table 3

Values of coefficients of thermal expansion α of $NaClO_3$ obtained by different authors

Temp. (°C)	$\alpha \times 10^6$/°C from		
	X-ray method		interferometric method
	present	DESHPANDE, MUDHOLKER	SHARMA
30	39.1	41.3	
40	40.4	41.8	45.3
60	43.2	43.3	46.6
80	46.0	45.3	48.0
100	48.8	47.7	49.5
120	51.5	50.6	51.0
140	54.3	54.0	52.6
160	57.1	57.9	54.2
180	59.8	62.2	55.9
200	62.6	67.1	57.6
220	65.4	72.3	59.4

constant of $NaBrO_3$ is larger than that of $NaClO_3$. Thus, the correlation between the thermal expansion and alkali-halogen distances is similar in the sodium halates as well as the potassium halides although the correlation between thermal expansion und lattice constant is different.

Unfortunately, not much information is available about the sodium-halogen distance in the mixed crystals of $NaClO_3$–$NaBrO_3$. There is only one report (SANJEEVI RAJA et al.) on the structure determination for a single composition in this series. From the data on atomic positions, a value of 3.88 Å is obtained for the Na-halogen distance in the mixed crystal $Na(ClO_3)_{0.7}\,(BrO_3)_{0.3}$. Based on the Na-halogen distances for the pure crystals given in the preceding para, a Vegard's rule calculation gives a value 3.93 Å for the Na-halogen distance in this mixed crystal. This is greater than the experimental value. Thus, apparently, in this series, the Na-halogen distance varies with composition with negative deviations from linearity unlike that in the

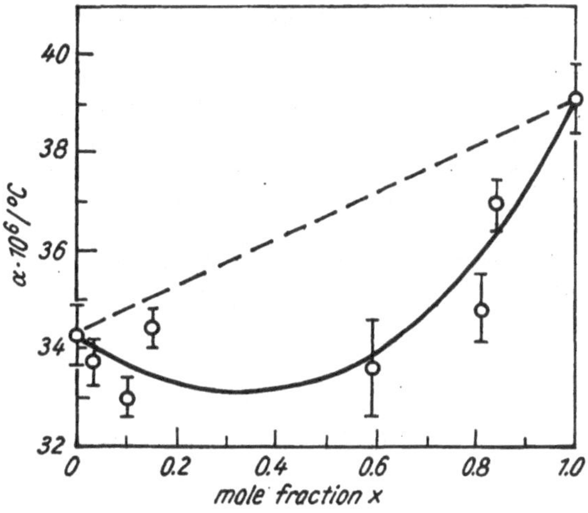

Fig. 4. Composition dependence of the coefficient of thermal expansion α at room temperature for the $Na(ClO_3)_x(BrO_3)_{1-x}$ system. Dashed line represents the variation expected from the additivity rule

Table 4
Values of coefficients of thermal expansion α of $NaBrO_3$ obtained by different authors

Temp. (°C)	$\alpha \times 10^6/°C$ from			
	X-ray method		interferometric method	
	present	WATHORE, KULKARNI	SHARMA	GANESAN*)
30	34.3	31.7		36.5
40	34.9	32.3	39.7	37.4
60	36.2	33.5	40.3	39.2
80	37.4	34.8	41.0	41.1
100	38.7	36.2	41.7	43.0
120	40.0	37.7	42.5	44.9
140	41.2	39.2	43.4	46.9
160	42.5	40.9	44.3	48.8
180	43.7	42.7	45.3	
200	45.0	44.5	46.4	
220	46.2	46.5	47.6	
240	47.5	48.5	48.8	
260	48.8	50.7	50.0	
280	50.0	52.9	51,4	
300	51.3	55.2	52.8	

*) interpolated from data given by GANESAN

KCl–KBr series. This difference in the composition dependence of the alkali halogen distances in the two series may account for the difference in the composition dependence of thermal expansion. However, a more systematic determination of the sodium-halogen distance in the sodium chlorate-sodium bromate series may throw more light not only on the composition dependence of thermal expansion but also that of several other physical properties of this series.

4. Conclusions

Lattice constants have been determined for several crystals in the mixed crystal series $Na(ClO_3)_x(BrO_3)_{1-x}$ in the temperature range room temperature to 320 °C. The lattice constants at room temperature are found to show large positive deviations from Vegard's rule. The linear coefficient of thermal expansion depends upon the composition in a non-linear fashion with negative deviations from linearity.

The authors are grateful to Prof. D. B. Sirdeshmukh for encouragement and helpful suggestions.

References

Abrahams, S. C., Bernstein, J. L.: Acta Cryst. B **33**, 3601 (1977)
Chandrasekhar, S., Madhava, M. S.: Acta Cryst. **23**, 911 (1967)
Chandrasekaran, K. S., Mohanlal, S. K.: Pramana **7**, 152 (1976)
Cohen, M. U.: Rev. Sci. Instr. **6**, 69 (1935)
Deshpande, V. T., Mudholker, V. M.: Acta Cryst. **13**, 483 (1960)
Ganesan, S.: J. Ind. Inst. Sci. **41**, 9 (1959)
Gluyas, M., Hunter, R., James, B. W.: J. Phys. D. 8, 1 (1975)
Heitala, J.: Ann. Acad. Sci. Fenn. AVI, No. 122 (1963)
Ivankina, M. S., Pozdeeva, E. V.: Izv. Vyss. Uch. Zav Fiz. **16**, 144 (1973)
Kantola, M.: J. State Inst. Tech. Res. Finland **3**, (1947)
Kleber, W., Koltermann, M.: Wiss. Z. Humboldt-Universität zu Berlin, p. 625 (1958)
Mason, W. P.: Phys. Rev. **70**, 529 (1946)
Mohanlal, S. K.: Ph. D. Thesis, Madurai University, Madurai, India (1971)
Nasr-Eddine, M., Sayetat, F., Vargas, H.: Bull. Soc. fr. Mineral Cryst. **98**, 378 (1975)
Pozdeeva, E. V., Avericheva, V. E., Guzeev, M. N.: VINITI p. 1564 (1975)
Radha, V., Gopal, E. S. R.: J. Ind. Inst. Sci. **50**, 26 (1968)
Salimaki, K. E.: Ann. Acad. Sci. Fenn. AVI, No. 56 (1960)
Sanjeevi Raja, C., Mohanlal, S. K., Chandrasekaran, K. S.: Z. Krist. **166**, 121 (1984)
Sharma, S. S.: Proc. Ind. Acad. Sci. A. **31**, 83 (1950)
Sirdeshmukh, D. B., Deshpande, V. T.: Proc. Ind. Nat. Sci. Acad. **38**, 167 (1972)
Sirdeshmukh, L., Reddy, Y. R.: Ferroelectrics **64**, 247 (1985)
Sirdeshmukh, D. B., Srinivas, K.: J. Mat. Sci. **21**, 4117 (1986)
Slagle, O. D., McKinstry, H. A.: Acta Cryst. **21**, 1013 (1966)
Srinivasan, K. R., Gopal, E. S. R.: Solid State Comm. **17**, 1119 (1975)
Subhadra, K. G., Hussain, K. A.: Cryst. Res. Technol. **22**, 827 (1987)
Wathore, T. N., Kulkarni, R. G.: J. Mat. Sci. **13**, 1817 (1978)

(Received February 10, 1987)

Authors' address:

Dr. (Miss) K. G. Subhadra, K. A. Hussain
Department of Physics, Kakatiya University
Warangal — 506009, India

Cryst. Res. Technol.	22	1987	9	1173–1178

H. Sobotta, H. Neumann, S. Mitaray, V. Riede, G. Kühn

Sektion Physik und Sektion Chemie, Wissenschaftsbereich Kristallographie, der Karl-Marx-Universität Leipzig

Infrared Optical Properties of $Cu_{1-x}Li_xInSe_2$ Thin Films

Infrared absorption spectra of $Cu_{1-x}Li_xInSe_2$ thin films are measured at room temperature in the wavenumber range from 100 to 600 cm^{-1}. The spectra exhibit two absorption bands in the wavenumber ranges 200–210 cm^{-1} and 330–355 cm^{-1} which are ascribed to vibrational modes due to In–Se and Li–Se vibrations, respectively. The influence of the phase transition from the chalcopyrite structure to the β-$NaFeO_2$ structure in the composition range $x = 0.5-0.6$ on the vibrational characteristics is established and discussed.

Infrarotabsorptionsspektren von dünnen $Cu_{1-x}Li_xInSe_2$-Schichten wurden bei Raumtemperatur im Wellenzahlbereich von 100 bis 600 cm^{-1} gemessen. Die Spektren zeigen zwei Absorptionsbanden in den Wellenzahlbereichen 200–210 cm^{-1} and 330–355 cm^{-1}, die Schwingungsmoden infolge von In–Se- bzw. Li–Se-Schwingungen zugeschrieben werden. Der Einfluß des Phasenübergangs von der Chalkopyritstruktur zur β-$NaFeO_2$-Struktur im Zusammensetzungsbereich $x = 0.5-0.6$ auf die Gitterschwingungseigenschaften wird nachgewiesen und diskutiert.

1. Introduction

Recently, the growth, structure and near-edge optical properties of $Cu_{1-x}Li_xInSe_2$ thin films have been reported (Mitaray et al.). It has been found that the films crystallize in the chalcopyrite structure at $x \leqq 0.5$ and in the orthorhombic β-$NaFeO_2$ structure at $x \geqq 0.6$. Furthermore, it has been established that the fundamantal absorption edge of the $Cu_{1-x}Li_xInSe_2$ mixed crystals is influenced by the structural phase transition and is due to direct allowed transitions in the chalcopyrite phase and due to indirect allowed transitions in the orthorhombic phase. This behaviour corresponds to the nature of the fundamental edge of the constituent compounds $CuInSe_2$ (Hörig et al. 1978) and $LiInSe_2$ (Hörig et al. 1984) of the mixed crystal system considered. It was the aim of the present work to get some first information about the lattice vibrational properties of the $Cu_{1-x}Li_xInSe_2$ mixed crystals by measuring and analysing the infrared absorption spectra of very thin films. In particular, it was of interest to investigate the influence of the structural phase transition on the vibrational characteristics.

2. Experimental

Powdered samples of $Cu_{1-x}Li_xInSe_2$ over the whole composition range were prepared by solid state reaction of corresponding mixtures of the prereacted compounds $CuInSe_2$ and $LiInSe_2$ and were used as starting material in the thin film growth experiments. Since both $CuInSe_2$ and $LiInSe_2$ evaporate incongruently (Hönle, Kühn) the flash evaporation technique was used to deposit the $Cu_{1-x}Li_xInSe_2$ thin films. High-resistivity single crystals of silicon were used as substrates because they exhibit no remarkable optical absorption in the wavenumber range below about 500 cm^{-1} (Ramdall, Rawcliffe) where phonon induced structures are expected in the optical spectra (Sobotta et al.). For the measurements a series of $Cu_{1-x}Li_xInSe_2$ thin films with x values varying between $x = 0$ and $x = 1$ in steps of $\Delta x = 0.1$ was prepared. The structure of the films was determined by RHEED measurements. It was found that in complete analogy to the results observed for $Cu_{1-x}Li_xInSe_2$ thin films on GaAs and glass substrates (Mitaray et al.) the films have the chalcopyrite structure at $x \leqq 0.5$ and the β-$NaFeO_2$ structure at $x \geqq 0.6$. All the films investigated were polycrystalline.

The thickness of the films was sochen in the range from 100 to 300 nm. Then, the frequencies of the maxima of the absorption bands due to infrared active vibrational modes are practically equal to the frequencies $\bar{\nu}_{TO}$ of the corresponding tranverse optical modes (Born, Huang) which facilitates the interpretation of the experimental data. The optical transmission spectra of the samples were measured at room temperature in the wavenumber range $\bar{\nu} = 100-600$ cm^{-1} at normal incidence of radiation. The absorption coefficient spectra of the films were determined from the measured transmittance of the film/substrate systems in the same manner as described in detail by Neumann et al. (1979).

3. Results and discussion

In agreement with previous measurements (Neumann et al. 1983) the absorption spectra of the $CuInSe_2$ thin films showed a well pronounced double-structured band in the wavenumber range $\bar{\nu} = 200-220$ cm^{-1} with two peak frequencies of 206 and 213 cm^{-1} and a rather weak second absorption band near 170 cm^{-1}. According to Neumann (1985a) these two bands are essentially due to the intense sphalerite-like modes in the lattice vibration spectrum of $CuInSe_2$ which are caused by In–Se and Cu–Se vibrations, respectively. The frequencies of these modes are given by (Neumann 1985a)

$$\bar{\nu}^2_{TO}(ASe) = \frac{\alpha(ASe)}{\pi^2 c^2} \left(\frac{1}{m_A} + \frac{1}{m_{Se}} \right), \tag{1}$$

where the m_i are the masses of the atoms participating in the vibration, c is the vacuum velocity of light, and A = Cu, In. The α(ASe) are the respective effective interatomic force constants which include contributions from bond-stretching, bond-bending and long-range Coulomb forces. An analysis of the lattice vibrational properties of elemental, binary and ternary semiconductors with tetrahedral coordination has shown that the α(ASe) depend only on the bond length d(ASe) and the individual bond ionicity f_i(ASe) of the A–Se bond (Neumann 1985a, b, 1987). In the case of $CuInSe_2$ the α(ASe) of relation (1) are found to be $\alpha(CuSe) = 17.0$ Nm^{-1} and $\alpha(InSe) = 31.6$ Nm^{-1}.

The spectra of the $Cu_{1-x}Li_xInSe_2$ thin films exhibit always two absorption bands near 210 cm^{-1} and in the range 330–355 cm^{-1} as shown in Figure 1. The variation with mixed crystal composition of the peak frequency of these bands is represented in Figure 2. The absorption band near 170 cm^{-1} due to the Cu–Se vibrations could not be detected in the mixed crystal spectra.

From Figure 2 we see that the peak frequency of the low-energy absorption band is obviously not influenced by the structural phase transition and remains nearly unchanged over the whole composition range. This result suggests the conclusion that this band can be ascribed to vibrations of the In atoms against the Se atoms with a transverse optical mode frequency again given by relation (1) with a practically unchanged force constant α(InSe). This result is not unexpected for the following reasons

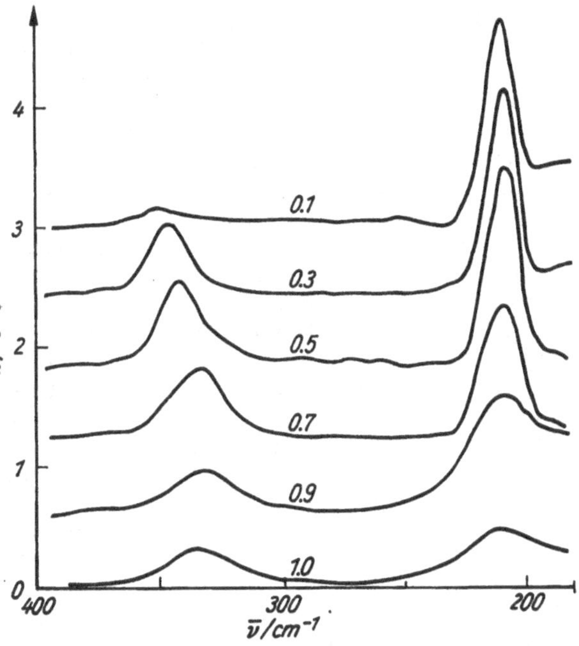

Fig. 1. Absorption spectra of $Cu_{1-x}Li_xInSe_2$ thin films with the molar fraction x as parameter. For clarity, with decreasing x the spectra are shifted against each other by $\Delta\alpha = 6 \cdot 10^3\ cm^{-1}$

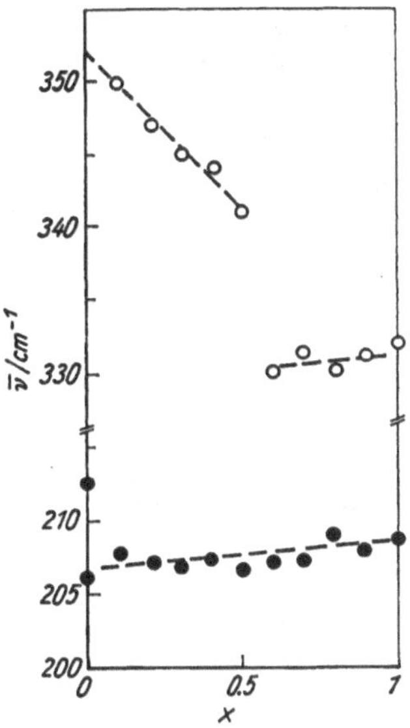

Fig. 2. Composition dependence of the peak frequencies of the two absorption bands shown in Figure 1

Firstly, it follows from model considerations for the lattice vibrations in β-$NaFeO_2$ structure crystals (NEUMANN 1986b) that a sphalerite-like mode with a frequency given by relation (1) exists in these compounds, too. Secondly, the In–Se bond length has a value of $d(InSe) = 0.2586$ nm in $CuInSe_2$ (PARKES et al.) while the mean value of the four In–Se bond lengths in $LiInSe_2$ is equal to $d(InSe) = 0.2569$ nm (HÖNLE et al.) which is only slightly below the bond length in $CuInSe_2$. Thirdly, the ionicity of the In–Se bond has the same value of $f_i(InSe) = 0.60$ in both compounds (NEUMANN 1983, 1986a). Thus, accounting for the d(InSe) and f_i(InSe) we expect that the α(InSe) and, therefore, also the $\bar{\nu}_{TO}$(InSe) (see relation (1)) increase slightly if we go from $CuInSe_2$ to $LiInSe_2$ which corresponds to the experimental findings (Fig. 2).

The second absorption band in the wavenumber range $\bar{\nu} = 330-355\ cm^{-1}$ was only observed in the mixed crystals and in $LiInSe_2$ and, therefore, can be ascribed to lattice vibrations caused by the Li atoms. From Figure 2 it follows that the peak frequency of this band decreases linearly with increasing Li content up to $x = 0.5$, then undergoes a sudden change from about 340 cm^{-1} to about 330 cm^{-1} between $x = 0.5$ and $x = 0.6$, and remains nearly constant in the composition range $x = 0.6-1.0$. Extrapolation to the dilute limit $x \to 0$ of the Li concentration yields a frequency of $\bar{\nu}_{loc}(Li) = 352\ cm^{-1}$ for the localized vibrational mode of the Li atoms in the $CuInSe_2$ lattice.

Obviously, the abrupt change in the peak frequency of this absorption band observed in the range $x = 0.5-0.6$ is related to the phase transition from the tetragonal chalcopyrite structure to the orthorhombic β-$NaFeO_2$ structure of the films which occurs in the same composition range. There may be two reasons for the observed discontinuity in the peak frequency. Firstly, it follows from NEUMANN (1985a) and from lattice vibration studies in the chalcopyrite-structure compound $LiInTe_2$ (KÜHN et al.) that in the chalcopyrite phase of the $Cu_{1-x}Li_xInSe_2$ mixed crystals the high-energy absorption band should be mainly due to a sphalerite-like mode caused by Li–Se vibrations with a transverse optical mode frequency $\bar{\nu}_{TO}$(LiSe) given by relation (1). The decrease in the peak frequency with increasing Li content up to $x = 0.5$ can certainly be ascribed to an increase in the Li–Se bond length d(LiSe) which gives rise to a decrease in the force constant α(LiSe) and, consequently, also in the frequency $\bar{\nu}_{TO}$(LiSe). Indeed, in the dilute limit $x \to 0$ it should be expected that $d(\text{LiSe}) \approx \approx d(\text{CuSe}) = 0.2484$ nm (PARKES et al.) whereas the average Li–Se bond length in $LiInSe_2$ is distinctly larger and equal to $d(\text{LiSe}) = 0.2538$ nm (HÖNLE et al.). Now, if we assume that this simple sphalerite-like mode model for the high-energy absorption band remains also valid for the orthorhombic phase of the mixed crystals at $x \geqq 0.6$ it is obvious that the sudden drop in the peak frequency and, thus, also in $\bar{\nu}_{TO}$(LiSe) between $x = 0.5$ and $x = 0.6$ can only be explained by an abrupt increase in the force constant α(LiSe) and, therefore, in the average bond length d(LiSe) at the structural phase transition. Furthermore, d(LiSe) should remain practically unchanged at $x \geqq 0.6$ because of the nearly constant peak frequency in this composition range (Fig. 2).

Secondly, it follows from group theory (NEUMANN 1986b) that the number of phonon branches in the β-$NaFeO_2$ lattice is twice as large as in the chalcopyrite lattice. Accounting for this fact it cannot be excluded that the phase transition from the chalcopyrite structure to the β-$NaFeO_2$ structure gives rise to the occurrence of additional zone-centre infrared active optical modes with frequencies just below the sphalerite-like mode due to Li–Se vibrations. Such modes have been, for instance, observed in the lattice vibrational studies of the compounds $LiInS_2$ (SOBOTTA et al.) and $LiGaO_2$ (NEUMANN et al. 1986) which also crystallize in the β-$NaFeO_2$ structure. In this case, the high-energy absorption band contains contributions from several vibrational modes, and the peak frequency of the band is not equal to the frequency $\bar{\nu}_{TO}$(LiSe) of the sphalerite-like mode given by relation (1) but represents an average value of the transverse optical mode frequencies of all infrared active vibrations in this wavenumber range which essentially depends on the oscillator strengths of the modes.

Obviously, to decide between the two possible reasons for the discontinuous behaviour of the peak frequency of the high-energy absorption band in the range of the structural phase transition, both a complete structural analysis of the whole mixed crystal system and a detailed infrared reflectivity analysis of the vibrational properties on bulk mixed crystals is required. Unfortunately, no experimental results concerning these problems have been reported in the literature until now.

So far, we have only considered the composition dependence of the peak frequencies of the two characteristic absorption bands with the result that the energetic position of the high-energy band is influenced by the structural phase transition while that of the low-energy band is not (Fig. 2). Looking at the absorption spectra shown in Figure 1 it is evident that the peak height of the absorption bands is lower in $LiInSe_2$ than, for instance, in $Cu_{0.5}Li_{0.5}InSe_2$. To follow this effect in more detail we have calculated the integrated absorption

$$A = \int \alpha(\bar{\nu})\, d\bar{\nu} \tag{2}$$

for both bands and plotted as a function of x in Figure 3. We see that the integrated absorption of both bands shows an approximately linear increase with x up to $x = 0.6$. At higher x values a marked decrease in both integrated absorptions occurs up to $x = 0.8$ while no essential changes are observed in the range $x = 0.8-1.0$. Obviously,

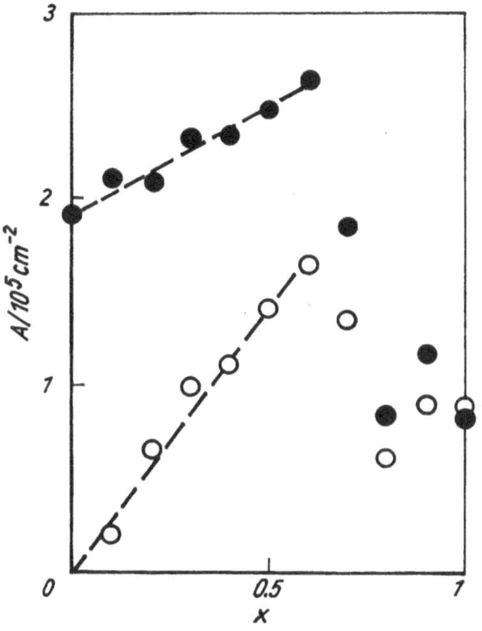

Fig. 3. Composition dependence of the integrated absorption of the low-energy (●) and high-energy absorption band (○) of Figure 1

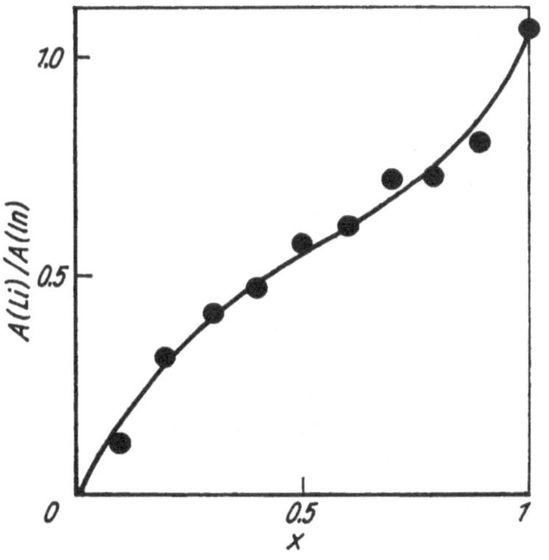

Fig. 4. Measured (points) and calculated with relation (4) (full curve) integrated absorption ratio as a function of the molar fraction x

this effect is also related to the structural phase transition, but no explanation of this phenomenon can be given at present. Finally, Figure 4 shows the composition dependence of the ratio A(Li)/A(In) of the integrated absorptions of the Li-induced and In-induced vibrational bands. As expected we have A(Li)/A(In) = 0 in $CuInSe_2$ and the maximum value of A(Li)/A(In) ≈ 1.1 in $LiInSe_2$. However, it is interesting to note that in spite of the structural phase transition in the range $x = 0.5-0.6$ the composition dependence of A(Li)/A(In) is continuous and can be described by the function

$$A(Li)/A(In) = 1.996x - 2.616x^2 + 1.648x^3 \quad (3)$$

(full curve in Figure 4). Relation (3) offers a simple method to determine the composition of $Cu_{1-x}Li_xInSe_2$ thin films from the ratio A(Li)/A(In) which can be easily measured.

References

BORN, M., HUANG, K.: Dynamical Theory of Crystal Lattices, Oxford 1954

HÖNLE, W., KÜHN, G.: J. Thermal Anal. **31**, 589 (1986)

HÖNLE, W., KÜHN, G., NEUMANN, H.: Z. anorg. allg. Chem. **543**, 161 (1986)

HÖRIG, W., NEUMANN, H., KÜHN, G.: phys. stat. sol. (b) **121**, K55 (1984)

HÖRIG, W., NEUMANN, H., SOBOTTA, H., SCHUMANN, B., KÜHN, G.: Thin Solid Films **48**, 67 (1978)

KÜHN, G., SCHUMANN, B., OPPERMANN, D., NEUMANN, H., SOBOTTA, H.: Z. anorg. allg. Chem. **531**, 61 (1985)

MITARAY, S., KÜHN, G., SCHUMANN, B., TEMPEL, A., HÖRIG, W., NEUMANN, H.: Thin Solid Films **135**, 251 (1986)

NEUMANN, H.: Cryst. Res. Technol. **18**, 1299 (1983)

NEUMANN, H.: Helv. Phys. Acta **58**, 337 (1985a)
NEUMANN, H.: Cryst. Res. Technol. **20**, 773 (1985b)
NEUMANN, H.: Cryst. Res. Technol. **21**, 1207 (1986a)
NEUMANN, H.: Cryst. Res. Technol. **21**, 1361 (1986b)
NEUMANN, H.: Cryst. Res. Technol. **22**, 99 (1987)
NEUMANN, H., HÖRIG, W., RECCIUS, E., SOBOTTA, H., SCHUMANN, B., KÜHN, G.: Thin Solid Films **61**, 13 (1979)
NEUMANN, H., KISSINGER, W., NEELS, H., RIEDE, V., SOBOTTA, H.: Cryst. Res. Technol. **21**, 925 (1986)
NEUMANN, H., SOBOTTA, H., RIEDE, V., SCHUMANN, B., KÜHN, G.: Cryst. Res. Technol. **18**, K90 (1983)
PARKES, J., TOMLINSON, R. D., HAMPSHIRE, M. J.: J. Appl. Cryst. **6**, 414 (1973)
RAMDALL, C. M., RAWCLIFFE, R. D.: Appl. Opt. **6**, 1889 (1967)
SOBOTTA, H., NEUMANN, H., RIEDE, V., KÜHN, G.: Cryst. Res. Technol. **21**, 1367 (1986)

(Received March 3, 1987)

Authors' addresses:

Dr. H. SOBOTTA, Dr. H. NEUMANN, Dr. V. RIEDE
Sektion Physik, Karl-Marx-Universität
Linnéstraße 5
Leipzig
DDR-7010

Prof. Dr. G. KÜHN, Dr. S. MITARAY
Sektion Chemie (WB Kristallographie), Karl-Marx-Universität
Scharnhorststraße 20
Leipzig
DDR-7030

Cryst. Res. Technol.	22	1987	9	1179–1182

S. BHUSHAN, L. C. GIRIYA

Department of Physics, Ravishankar University, Raipur, India

Photoconductivity of (ZnS, CdSe): Dy

Photoconductivity rise and decay curves have been investigated for (ZnS, CdSe): Dy as a function of flux concentration, concentration of Dy, applied voltage, atmosphere during preparation and different excitation intensities. Substitution of Dy increases the photocurrent by five times. Maximum photocurrent is also observed at a particular concentration of flux. The decay is of hyperbolic nature and the trap depths are found at depths 0.49 eV and 0.54 eV which depend upon the concentration of Dy.

Introduction

II–VI compounds form a very important class of photoconductors and are well studied. These compounds can be used in mixed bases form to produce maximum absorption in visible region. Photoconductivity of rare earth doped mixed bases has not drawn any attention. Only CdS doped with Eu or Dy has been studied to some extent (FIALKOVSKAYA KHIRUNENKO). These studies include the ionization energy of the basic state of these impurities from the spectral studies of photoconductivity at 297 K and 77 K. The aim of the present paper has been to investigate the effect of rare earth doping on the mixed base (ZnS, CdSe), which has not been attempted so far. Mixing of ZnS was done to improve the photoresponse in the visible region.

Experimental

The samples have been prepared by appropriate mixing of the bases ZnS (Grade I from Johnson and Matthey Chemicals Ltd.) CdSe (Ultrapure from E. Merck), dysprosium oxide (99.9% Indian Rare Earth Ltd.) NaF (G.R., E. Merck) and then firing at 700 °C for 1 hour. Dy and NaF were mixed in solution form and the mixture was dried and thoroughly mixed in agate pestle-mortar. The photoconductivity cell was prepared by pressing the finely ground materials between a conducting glass plate of resistance = 200 Ohm/cm^2 and an aluminium plate. 60 W commercial bulb was used as excitation source and a nanometer (M/s Scientific Equipment Roorkee) for recording the photocurrent. Biasing was done by using a regulated power supply.

Results and discussions

The variation of dark current with applied voltage is shown in Figure 1. This plot shows a linear variation of dark current with the voltage applied. Figure 2 shows the rise and decay curves (ZnS, CdSe): Dy at different concentrations of the flux NaF. From these variations, it was concluded that the best concentration of flux was 0.026% by weight of the mixed base. This flux concentration was selected for preparing the samples with different concentrations of Dy, results of which are shown in Fi-

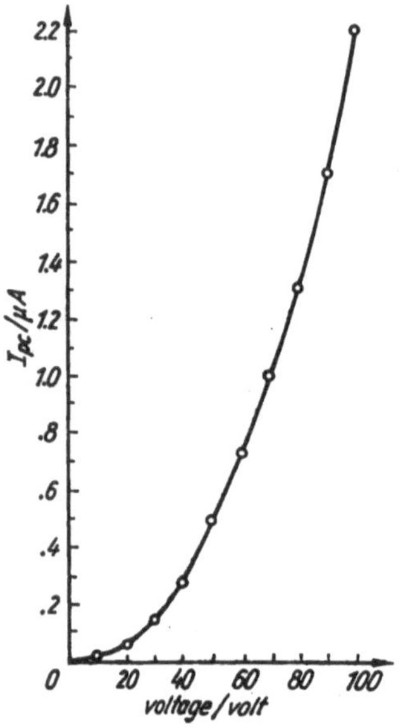

Fig. 1. Dependence of dark current on applied voltage for (CdSe, ZnS): Dy

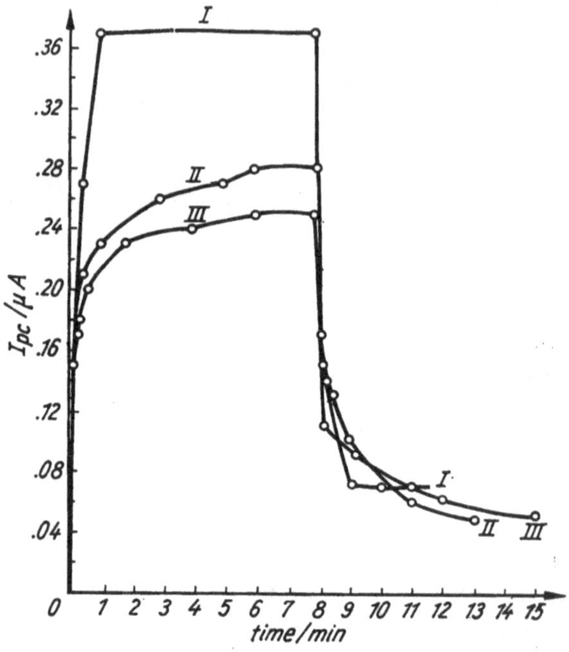

Fig. 2. Photoconducting rise and decay curves for (CdSe, ZnS): Dy at different flux concentrations. I (2 CdSe, 1 ZnS): Dy (.005%), flux (.052%), II (2 CdSe, 1 ZnS): Dy (.005%), flux (.026%), III (2 CdSe, 1 ZnS): Dy (.005%), flux (.104%)

gure 3. In this diagram results of (ZnS, CdSe) flux are also plotted. An increase of about five times is observed due to substitutions of Dy. (ZnS, CdSe) without flux or impurities showed high dark current and no increase due to excitations of light photons was observed. It was also attempted to prepare the Dy-doped samples without any use of flux but no success was made. As seen from this figure the best concentration of Dy giving maximum photocurrent has been 0.005% by weight Dy of mixed base (ZnS, CdSe). The ratio of ZnS and CdSe was decided on the basis of photoresponse which was found to be maximum at the ratio 2:1 at fixed concentration of flux and impurity. The role of impurity Dy may be considered in terms of release of valence electrons according as $Dy^{\circ} \rightarrow Dy^{+++} + 3e$. The lower values of photocurrent at lower and higher concentrations may be due to lesser release of electrons from Dy and due to concentration quenching, respectively. Plot of saturated photoresponse at different concentrations of Dy are shown in Figure 4. The rise and decay curves at different applied voltages are shown in Figure 5. Though the nature of these curves are the same, a higher increase in photocurrent is observed at 100 V. This shows that the photoelectrons are generated in sufficiency but at higher voltages more carriers reach the respective electrodes.

The photoresponse of the materials prepared in N_2 and air are shown in Figure 6. The photocurrent for samples prepared in N_2 are less. These samples show weak luminescent emmission under UV light and this can be reason for slightly lower photocurrent. The photoresponse at different intensities is similar and the saturated photoresponse (Fig. 7) shows a faster increase in the beginning and then is comparatively slower at high intensities.

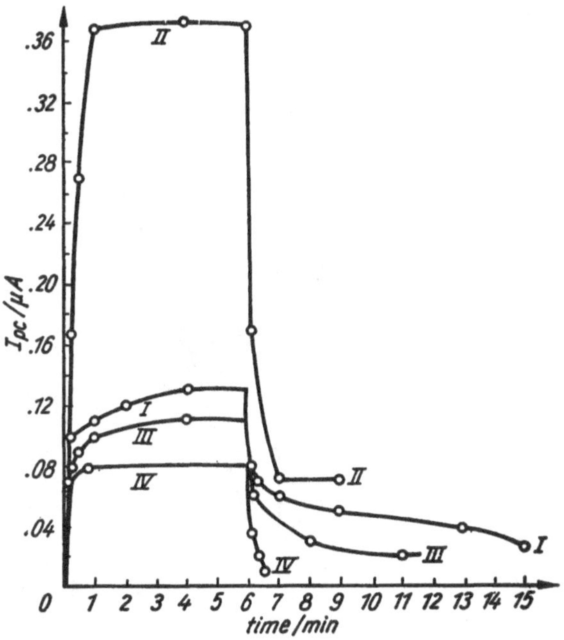

Fig. 3. Photoconducting rise and decay curves for (CdSe, ZnS): Dy at different concentrations of Dy. I (2 CdSe, 1 ZnS), flux (.052 %) Dy .010 %, II (2 CdSe, 1 ZnS), flux (.052 %) Dy .005 %, III (2 CdSe, 1 ZnS), flux (.052 %) Dy .002 %

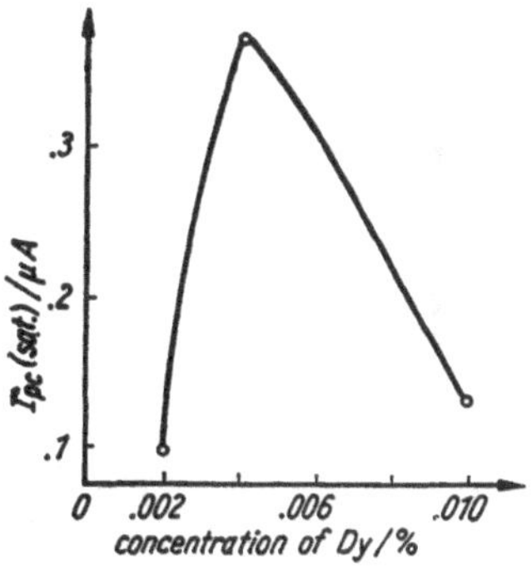

Fig. 4. Plot of saturated photocurrent at different concentrations of Dy

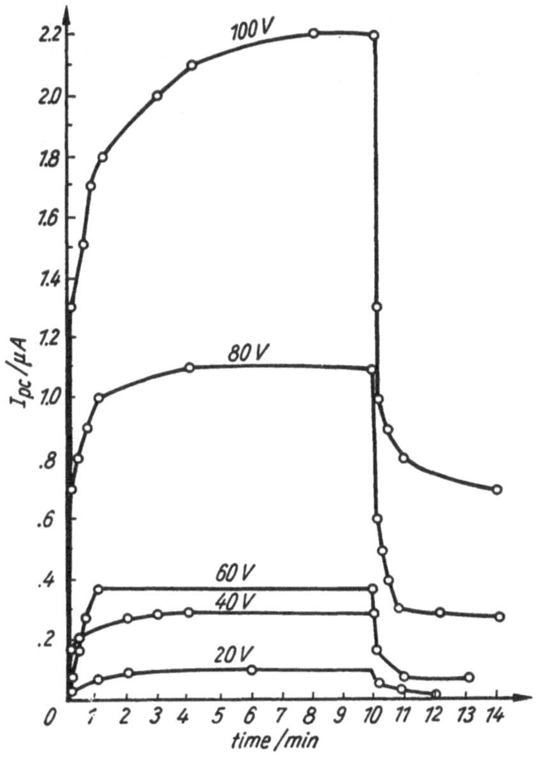

Fig. 5. Photoconducting rise and decay curves of (CdSe, ZnS): Dy at different applied voltages

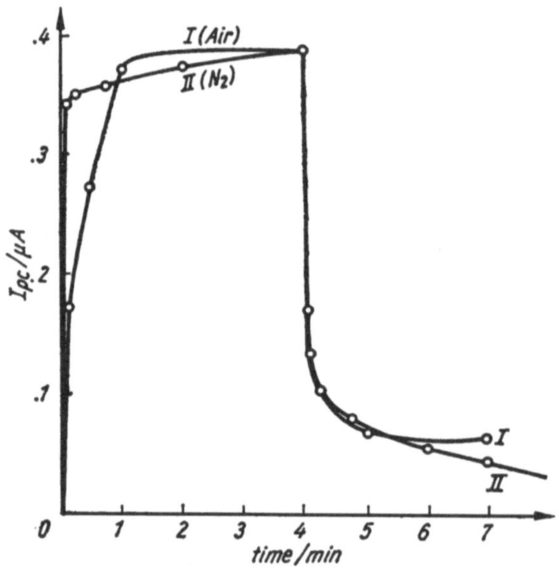

Fig. 6. Photoconducting rise and decay curves of (CdSe, ZnS): Dy prepared in different atmospheres

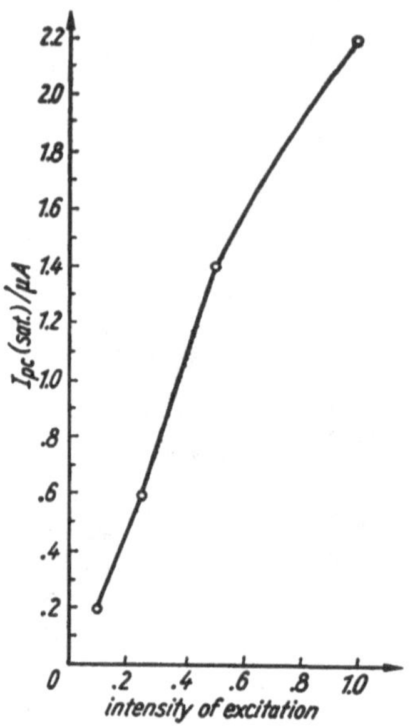

Fig. 7. Saturated photocurrents at different intensities for (CdSe, ZnS): Dy

From peeling off the decay curves into possible number of exponentials and using the relation $I = I_0 e^{-pt}$ and $p = S \cdot \exp(-E/kt)$ the trap depths were calculated. Here I represents the photocurrent at any instant t, I_0 the current at the moment light is interrupted, p the probability of escape of an electron from traps per second, E the trap depth and S the number per second that the quanta from crystal vibrations attempt to eject the electrons from traps multiplied by the transition from the trap to the conduction band. Two kinds of traps have been found with depths 0.54 eV and 0.49 eV and these are found to depend upon the concentration of Dy.

The authors thank Dr. R. K. Thakur for his encouragement throughout the work.

References

BHUSHAN, S., DIWAN, D.: Nat. Acad. Science Letter **7**, 371 (1984)
BHUSHAN, S., DIWAN, D.: J. Mat. Science Letter **5**, 723 (1986)
FIALKOVSKAYA, O. V., KHIRUNENKO, I. L.: Ukr. fiz. (USSR) **19**, 1533 (1974)
FIALKOVSKAYA, O. V., KHIRUNENKO, L. I.: Ukr. Fiz. Zh. (USSR) **19**, 1386 (1974)

(Received February 13, 1987)

Authors' address:
Dr. Sh. BHUSHAN, L. C. GIRIYA
Department of Physics, Ravishankar University
Raipur-492 010, India

Cryst. Res. Technol.	22	1987	9	1183–1187

Sh. Akchalov, A. Alybakov, M. Zhanibekov, T. K. Karashev, N. Tojchiev

Institute of Physics, Academy of Sciences of Kirghiz SSR;
Kirghiz State University; Kirghiz Women's Pedagogical Institute

Hyperfine Structure of Cr^{5+} EPR Spectra in X-irradiated Potassium Halide Crystals Doped with CrO_4^{2-} and Ca^{2+}

The hyperfine structure has been investigated in the EPR spectra of X-irradiated KCl: CrO_4^{2-}, Ca^{2+} and KBr: CrO_4^{2-}, Ca^{2+} crystals. Preliminarily spectra were simplified by means of heating the crystals up to 400 K for KCl and 440 K for KBr, that destroys the less stable $CrO_4^{3-} \cdot V_a^+$ centers. It is ascertained, that principal directions of g- and A-tensors do not coincide, and principal A-values are determined. Calculations, making use of the experimental meanings of g- and A-tensor components, showed, that the degree of distortion of oxygen tetrahedra in $CrO_4^{3-} \cdot Ca^{2+} \cdot V_a^+$ centers is almost the same in both KCl and KBr crystals; besides, these calculations suggest the existence of a strong covalent bonding between the central Cr^{5+} ion and four oxygen ligands in CrO_4^{3-} ions.

Исследована сверхтонкая структура в спектре ЭПР облученных кристаллов KCl:CrO_4^{2-}, Ca^{2+} и KBr:CrO_4^{2-}, Ca_a^+. Предварительно спектры упрощены нагреванием кристаллов до 400 K для KCl, 440 K для KBr, в результате которого разрушаются менее стабильные центры $CrO_4^{3-} \cdot V_a^+$. Установлено, что главные направления тензоров g и A не совпадают, определены главные значения тензора A. Вычисления с использованием экспериментальных значений компонент тензоров g и A показали, что степень искажения кислородных тетраэдров в центрах $CrO_4^{3-} \cdot Ca^{2+} \cdot V_a^+$ в кристаллах KCl и KBr почти одинакова и что существует сильная ковалентная связь между центральным ионом Cr^{5+} и четырьмя кислородными лигандами в ионе CrO_4^{3-}.

1. Introduction

At it was ascertained earlier (Alybakov et al.), X-irradiation of KCl: CrO_4^{2-}, Ca^{2+} and KBr: CrO_4^{2-}, Ca^{2+} crystals results in the creation of two types of paramagnetic centers, being $CrO_4^{3-} \cdot V_a^+$ and $CrO_4^{3-} \cdot Ca^{2+} \cdot V_a^+$ complexes. Centers of each type are situated in the crystal lattice in several magnetically inequivalent positions. As a result, EPR spectrum consists of a great number of lines overlapping with each other when one changes the sample orientation in the magnetic field. The complexity of the spectrum does not allow to study in detail the hyperfine structure (HFS) resulting from the interaction of unpaired electron spin with ^{53}Cr nucleus. At the same time it is ascertained, that afore-mentioned paramagnetic centers have the different thermal stability. That allows to simplify the spectrum considerably by means of thermal destruction of less stable $CrO_4^{3-} \cdot V_a^+$ centers. After such treatment one can study the HFS of the EPR spectrum of more stable $CrO_4^{3-} \cdot Ca^{2+} \cdot V_a^+$ centers. The results of such investigation are given in this paper.

2. Experimental

The studied crystals were grown in air by the Kyropoulos method in a platinum crucible from KCl and KBr of special purity doped with 0.5 mol.% of K_2CrO_4 and 0.05 mol.% of $CaCl_2$. X-irradiation for periods of up to 7 h was performed at the room temperature (tungsten target tube, 55 kV, 10 mA). To destroy $CrO_4^{3-} \cdot V_a^+$ centers, irradiated samples were heated up at the rate of 10 K/min to 400 K for KCl and 440 K for KBr. EPR spectra were recorded on a standard radiospectrometer RE-1307 at liquid nitrogen temperature.

3. Results and discussion

Since $CrO_4^{3-} \cdot Ca^{2+} \cdot V_a^+$ centers in KCl and KBr crystals occupy six magnetically inequivalent positions (Alybakov et al. 1982, 1984) their EPR spectrum at arbitrary crystal orientation is rather complicated. It gets simplified when magnetic field is oriented along the crystallographic axes (Fig. 1). Intensive lines belong to the even chromium isotopes and weak HFS lines are due to interaction of unpaired electron with nucleus of ^{53}Cr isotope. Principal g-values, as determined early (Alybakov et al. 1982, 1984), are as follows: in KCl crystals $g_x = 1.938$, $g_y = 1.948$, $g_z = 1.966$, in KBr crystals $g_x = 1.936$, $g_y = 1.945$, $g_z = 1.965$.

To study the angular variations of the separation between the HFS lines, we used the orthogonal axes connected with principal directions of g-tensor. Obtained dependences are shown in Figure 2. Principal values of hyperfine interaction tensor and direction cosines of its principal directions with respect to principal axes of g-tensor, as determined by Schonland method, are given in Table 1.

As it is seen from the Table 1, principal directions of g- and A-tensors do not coincide. That indicates the symmetry of Cr^{5+} surrounding in $CrO_4^{3-} \cdot Ca_a^{2+} \cdot V^+$ centers

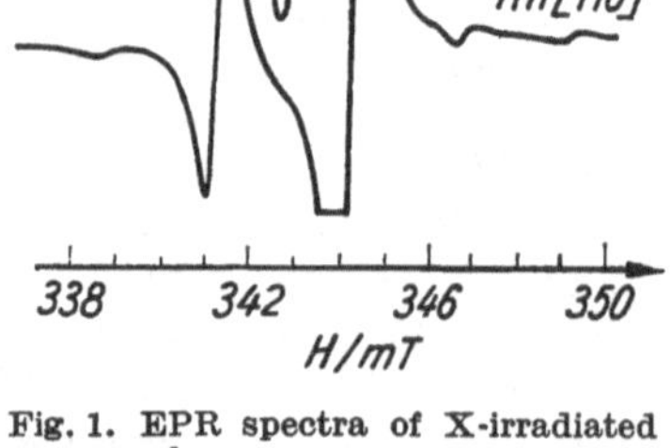

Fig. 1. EPR spectra of X-irradiated KCl: CrO_4^{2-}, Ca^{2+} crystal after heating up to 400 K for two orientations of the crystal in the magnetic field

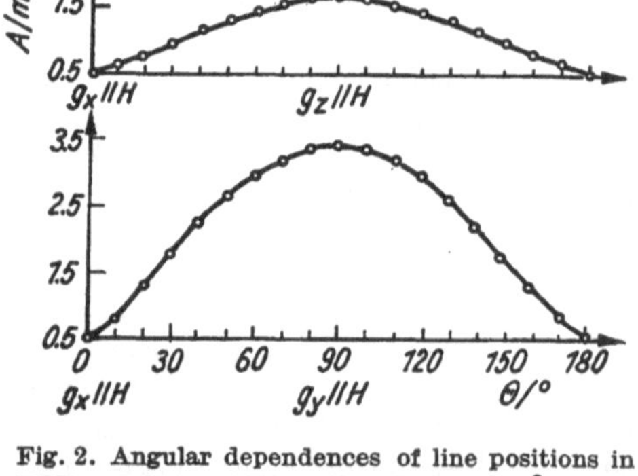

Fig. 2. Angular dependences of line positions in EPR spectrum of X-irradiated KBr: CrO_4^{2-}, Ca^{2+} crystal after heating up to 440 K

Table 1

Principal meanings and direction cosines of principal directions of hyperfine interaction tensor with respect to g-tensor axes for $CrO_4^{3-} \cdot Ca^{2+} \cdot V_a^+$ centers in KCl and KBr crystals

	Principal meanings of A-tensor, 10^{-4} cm^{-1}		direction cosines with respect to direction of		
	KCl	KBr	g_x	g_y	g_z
A_x	14.6	14.6	0	0	1
A_y	30.6	30.8	0	1	0
A_z	5.0	4.5	1	0	0

to be lower than rhombic (MEILMAN, SAMOILOVICH). These paramagnetic centers are formed as a result of the transition in a pentavalent state of chromium ions being in $CrO_4^{2-} \cdot Ca^{2+}$ complexes with C_{2v} symmetry. Results of HFS investigation suggest, that additional anion vacancy, compensating one more excess negative charge of CrO_4^{3-} ion, is situated at the place of the nearest anion in the [110] direction; in such a case symmetry of paramagnetic center is lowered to C_s. Model of such center is shown in Figure 3. In unperturbed CrO_4^{2-} ion Cr–O bonds are alligned along the [111] directions. In $CrO_4^{3-} \cdot Ca^{2+} \cdot V_a^+$ center CrO_4 tetrahedron is distorted because of Ca^{2+} and V_a^+ disposition in an immediate proximity, and Cr–O bonds may be deviated from [111] directions.

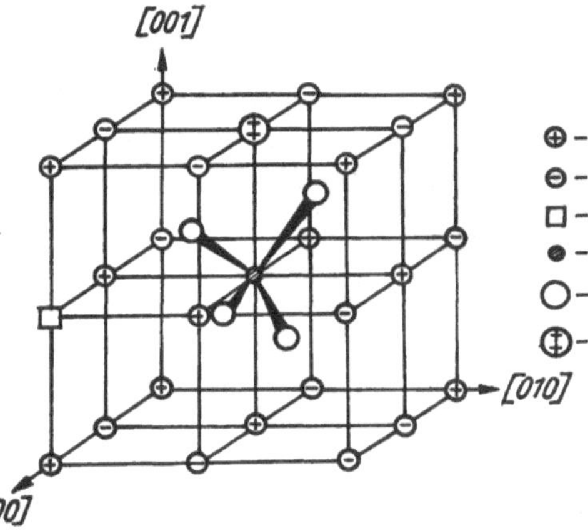

Fig. 3. Model of $CrO_4^{3-} \cdot Ca^{2+} \cdot V_a^+$ paramagnetic center

Chromium in CrO_4^{3-} ion is in the pentavalent state and is surrounded by four oxygens forming a tetrahedron. The ground state of Cr^{5+} is ^2D ($L = 2$, $S = 1/2$). In crystalline field of pure tetrahedral symmetry ^2D state splits in the doublet and triplet, doublet being the lowest in the energy. When the symmetry is lowered down to axial, doublet splits in the orbital states d_{z^2} and $d_{x^2-y^2}$. If the lowest (ground) state is d_{z^2}, one finds the relations $g_{\parallel} > g_{\perp}$, $A_{\parallel} < A_{\perp}$; for $d_{x^2-y^2}$ ground state the signs in these relations are changed (BANKS et al.; WERTZ, BOLTON). In axial symmetry $g_{\perp} = g_x = g_y$. In crystalline fields of lower symmetry d_{z^2} and $d_{x^2-y^2}$ orbitals are mixed, and $g_x \neq g_y$; however, above mentioned relations remain correct, with the difference that unpaired electron is localized on d_{z^2} orbital with some admixture of $d_{x^2-y^2}$, when $g_z > g_x, g_y$, $A_z < A_x, A_y$; when $g_z < g_x, g_y$ and $A_z > A_x, A_y$, the ground state is $d_{x^2-y^2}$ with some admixture of d_{z^2}. Obtained correlation between principal g- and A-values indicates, that unpaired electron is localised mainly on d_{z^2} orbital.

For paramagnetic centers with C_s symmetry following equations are correct (EZZEH, MCGARVEY):

$$g_{zz} = 2.0023 - 8b^2[\eta_2 + f^2(\eta_1 - \eta_2)]\,,$$
$$g_{xx} = 2.0023 - 2(\sqrt{3}a + b)^2\,[\eta_1 + f^2(\eta_2 - \eta_1)]\,,$$
$$g_{yy} = 2.0023 - 2\,(\sqrt{3}a - b)^2\,\eta_3\,,$$
$$A_{zz} = -K + P\{\tfrac{4}{7}\,(1 - 2b^2) + (\eta_3/7)\,(\sqrt{3}a - b)\,(\sqrt{3}a + 3b) + {}$$
$$+ \tfrac{1}{7}\,(\sqrt{3}a + b)\,(\sqrt{3}a - 3b)\,[\eta_1 + f^2(\eta_2 - \eta_1)] + (g_{zz} - 2.0023)\}\,,$$
$$A_{xx} = -K + P\{-\tfrac{2}{7}\,(1 - 2b^2) - (4\sqrt{3}/7)\,ab - (4\sqrt{3}/7)\,ab[\eta_2 + {}$$
$$+ f^2(\eta_1 - \eta_2)] - (\eta_3/7)\,(\sqrt{3}a - b)\,(\sqrt{3}a + 3b) + (g_{xx} - 2.0023)\}\,,$$
$$A_{yy} = -K + P\{-\tfrac{2}{7}\,(1 - 2b^2) + (4\sqrt{3}/7)\,ab + {}$$
$$+ (4\sqrt{3}/7)\,ab[\eta_2 + f^2(\eta_1 - \eta_2)] - \tfrac{1}{7}\,(\sqrt{3}a + b)\,(\sqrt{3}a - 3b) \times$$
$$\times\,[\eta_1 + f^2(\eta_2 - \eta_1)] + (g_{yy} - 2.0023)\}\,.$$

where

$$P = 2.0023 g_N \beta_e \beta_N \langle r^{-3}\rangle\,, \qquad a^2 = 1 - b^2\,,$$
$$\eta_1 = \lambda/\Delta E_{yz}\,, \qquad \eta_2 = \lambda/\Delta E_{xy}\,, \qquad \eta_3 = \lambda/\Delta E_{xz};$$

K is the isotropic Fermi contact term, β_e and β_N are the Bohr and nuclear magnetons, and λ is the spin-orbit coupling constant.

One can determine the parameters of these equations, making use of experimental values of g- and A-tensor components. Because of uncertainty of signs of A-tensor components these equations may have different solutions. Our analysis showed, that real values of b and P may be obtained only with positive meanings of A-tensor components. The results of the calculations, carried out with assumption $\eta_1 = \eta_3$, is given in Table 2. Value of $\chi = -\frac{3}{2}\varkappa\langle r^{-3}\rangle$ is also included in the table.

Table 2

Crystal	$\varkappa$	b^2	f^2	η_2	η_3	χ (a. u.)	P (10^{-4} cm^{-1})
KCl	0.61	0.1156	0.2758	0.0516	0.0070	−2.33	−25.73
KBr	0.58	0.1122	0.2664	0.0532	0.0074	−2.26	−26.21

The comparison of P with this parameter meaning for free Cr^{5+} ion, P_0 (according to MCGARVEY, $P_0 = -50 \cdot 10^{-4}$ cm^{-1}), shows that in both crystals P/P_0 is only slightly greater than 0.5. This indicates the existence of strong covalent bonding between the central Cr^{5+} ion and four oxygen ligands in CrO_4^{3-} ions. Obtained χ meanings are close to this parameter values for Cr^{5+} ion in other compounds (EZZEH, MCGARVEY; GREENBLATT, PIFER). The data of Table 2 allow to conclude, that in both KCl and KBr crystals degree of distortion of oxygen tetrahedra in $CrO_4^{3-} \cdot Ca^{2+} \cdot V_a^+$ centers is almost the same.

References

ALYBAKOV, A. A., TOJCHIEV, N.: phys. stat. sol. (b) **101** (No. 2) 597 (1980)
ALYBAKOV, A. A., TOJCHIEV, N., AKCHALOV, Sh.: phys. stat. sol. (b) **109** (No. 1) 295 (1982)
ALYBAKOV, A. A., TOJCHIEV, N.: Izv. AN Kirghiz SSR, No. 5, 18 (1985)

ALYBAKOV, A. A., GUBANOVA, V. A., ZHANIBEKOV, M., TOJCHIEV, N.: phys. stat. sol. (b), **121**, K205 (1984)
AKCHALOV, Sh., ZHANIBEKOV, M., TOJCHIEV, N.: in: Ionic and Electronic processes in Ionic Crystals, Theses of Reports, Osh 1986, p. 86—87 (in Russian)
BANKS, E., GREENBLATT, M., MCGARVEY, B. R.: J. Chem. Phys. **47** (No. 10) 3772 (1967)
EZZEH, C., MCGARVEY, B. R.: J. Chem. Phys. **61** (No. 7) 2675 (1974)
GREENBLATT, M., PIFER, J. H.: J. Chem. Phys. **70** (No. 1) 116 (1979)
MCGARVEY, B. R.: J. Phys. Chem. **71**, 51 (1967)
MEILMAN, M. L., SAMOILOVICH, M. I.: Introduction in EPR Spectroscopy of Doped Single Crystals, Moscow 1977 (in Russian)
WERTZ, J. E., BOLTON, J. R.: Theory and Practical Applications of EPR Method, Moscow 1975 (in Russian)

(Received February 17, 1987)

Authors' addresses:

Sh. AKCHALOV, Pr. A. A. ALYBAKOV, M. ZHANIBEKOV
Institute of Physics, Acad. Sci. of Kirghiz SSR
Leninsky Prospekt 265a
720071 Frunze, USSR

T. K. KARASHEV
Kirghiz State University
720024 Frunze, USSR

N. TOJCHIEV
Kirghiz Women's Pedagogical Institute
720040 Frunze, USSR

Book Review

J. Delmonte

Origins of Materials and Processes

Technomic Publishing Comp., Inc., Lancaster, Basel, 1985, 373 p., 189 figs., 568 ref., Cloth $ 30.00, ISBN 87 762-420-8

Wer das Wagnis unternimmt, ein Kompendium zu schreiben, das den Zeitraum vom Urknall bis zur Frühgeschichte überspannt, der muß weit gereist sein, fleißig gesammelt und sich einen umfassenden Ein- und Überblick verschafft haben. Die Herkunft der gebräuchlichen Materialien und Werkstoffe, die Gewinnungs- und Produktionsweisen sowie deren ständige Vervollkommnung beanspruchen eine möglichst lückenlose Darstellung. So ist es denn auch: Der Autor hat wenig besuchte Gebiete in aller Welt bereist u. a. die burmesisch-chinesische Grenze, die Stein-Wälder bei Kunming, bekannte Stätten in Europa, im Vorderen Orient und in Australien, hier die Höhlenkunst der Ureinwohner, der Aborigines studierend Das Resultat der jahrelangen Bemühungen ist umfassend und in sich konsistent: Man erfährt, woher die Materialien seinerzeit stammten, wie und von wem sie zuerst genutzt und in der Folgezeit weiterentwickelt wurden.

Das erste Kapitel, ,,Geologic Origins of Materials", stützt sich in seinen Aussagen wesentlich auf Verteilungsanalysen und Altersbestimmungen, berührt aber auch klimatische, atmosphärische und extraterrestrische Einflüsse sowie den Ursprung des Lebens. Im 2. Kapitel, ,,Precambrian to Prehistory", werden u. a. die Kontinentaldrift, die Änderungen des Erdmagnetfeldes, Probleme der Evolution, Höhlenfunde, Vertreter der frühen Menschheit, behandelt. Unter 3., ,,Calendrical Records", werden — ausgehend von astronomischen Konzepten und Referenzdaten für Kalendersysteme — die Kalender faktisch aller Kulturvölker vorgestellt. Den ,,Origins of Writing" ist das 4. Kapitel gewidmet. Es wird getragen von Darlegungen über die Ursprünge der bildlichen oder schriftlichen Verständigung. Stichworte wie Steinkunst, Piktogramme, Ideogramme, Hieroglyphen, älteste Zahlensymbole, Cuneiform-Schrift (sumerisch) bis hin zum griechischen und lateinischen Alphabet belegen die Spannweite dieses Kapitels. Die Materialien auf denen geschrieben wurde dürfen nicht fehlen: Diverse Gesteine, Knochen, Siegelmaterialien, Papyrus, Bambus, Wachstafeln, Stoffe Papier usw. wie sie 5., ,,Materials for Written Records", enthält. Das 6. Kapitel, ,,Ancient Metallurgical Practices", informiert über die frühgeschichtliche Gewinnung, Raffination und Verarbeitung der Edelmetalle, des Kupfers des Eisens, der Bronzen, in den bedeutendsten frühen Reichen. Den wohl ältesten ,,künstlichen" Werkstoffen, ,,Ancient Ceramics and Glasses", und den diesbezüglichen Entwicklungen in den verschiedenen Gebieten der Erde ist das 7. Kapitel gewidmet. Unter 8., ,,Natural Organic Fibres", rangieren Tierhäute, Flachs, Baumwolle, Seide, Wolle, weitere Textilien und auch Holzprodukte. Unter 9., ,,Miscellaneous Materials used by Early Man", werden vor allem Materialien (Gesteine) für Bauten, steinzeitliche Geräte, für Monumente, Statuen, Preziosen u. a. Kunstgegenstände erörtert.

Diese knapp, jedoch umfassend angelegte ,,Mini-Enzyklopedie" wendet sich an alle allgemein an der Sache Interessierten, aber auch an Spezialisten, wie Archäologen, Anthropologen, Geologen, Mineralogen, Metallurgen, Keramiker, Textil- und Glasfachleute. Dem Buch ist weite Verbreitung und eine deutsche Ausgabe zu wünschen.

D. Schulze

Cryst. Res. Technol.	22	1987	9	1189–1192

R. ASSENOV, N. V. IZMAILOV, V. A. MOSHNIKOV, N. P. YAROSLAVTSEV, D. A. YASKOV

The Leningrad Electrotechnical Institute, USSR, and
Semiconductor Physics and Technological Research Laboratory, University of Sofia, Bulgaria

Investigation of the Acoustic Attenuation in Lead and Tin Tellurides Obtained by the Iodide Method

The products of the reaction for obtaining lead telluride and tin telluride, synthesized by the iodide method, were analyzed by the method of acoustic attenuation (internal friction). The results of the investigations are in agreement with the model, suggested earlier, of self-alloying of tellurides by metallic component.

Методом акустического затухания (внутреннего трения) проанализированы продукты реакции получения теллурида свинца и теллурида олова, синтезированных по подидной методике. Результаты исследований согласуются с ранее предложенной моделью самолегирования теллуридов металлической компонентной.

The iodide method is one of the promising methods of obtaining chalcogenides (MOLDOVANOVA et al.). KWESTROO has described in detail the technological conditions of SnTe synthesis in SnI_2 solvent, and the principal advantages of the iodide method: lower temperature of sintering (appr. 400 °C), limited access to oxygen, purification of foreign admixtures, and improvement in the electrophysical characteristics of the material. Moreover, SnTe is obtained with sufficiently low concentrations of carriers of the p-type charge, corresponding as a rule to the metallic boundary of the unalloyed material. Analogous technological conditions in the synthesis of PbTe secure the obtaining of samples of n-type conductivity.

ASSENOV et al. have examined a model describing the observed changes in the electrophysical properties by self-alloying of the material in the process of its obtaining and treatment. They have also analyzed the products of the reaction of obtaining lead telluride by the low-temperature iodide method by means of X-ray diffractiometry. It was established that when dissolving Sn and Te in SnI_2 there occurs the formation of SnTe at temperatures of about 450 °C. Further on during the annealing of the samples obtained in a flow of inert gas there appears separation of iodides and the material obtained is SnI_4.

Nevertheless, a SnO_2 phase was established in the reaction product. It may be assumed that the initial material is enriched in metal in the process of treatment, and its excess is oxidized. However, it was not possible to establish excess of tin in the SnTe samples by direct methods. The attempts to establish phases containing lead in PbTe or tin in SnTe by means of X-ray diffractometry, and also by X-ray spectral microanalysis, yielded no results.

This shows that if a second phase in the form of pure elements exists in the tellurides obtained by the iodide method, their content does not exceed the sensitivity limit of X-ray diffractometry. Besides that, the negative results obtained from the X-ray spectral microanalysis of the micro-inclusions is an evidence of the fact that even if the individual components do exist as microseparations of lead (tin), their concentration and size is extremely small.

The aim of the present work was to analyze the samples of SnTe and PbTe obtained by the iodide method by a procedure based on the method of acoustic attenuation or internal friction (MITROKHIN et al.). This method has been successfully used earlier for analyzing foreign microinclusions in silicon (IZMAILOV et al.) and in binary compounds A^3B^5 (GLUSHKOV et al.), grown by the traditional methods of semiconductor technology.

The essence of the method uses lies in that, upon taking the temperature dependence of the acoustic attenuation in samples from the material containing inclusions of the second phase, at the melting point of these inclusions there appear peaks from whose size it is possible to evaluate also the volume concentration of the second phase. The peaks of internal friction conditioned due to the absorption of mechanical energy at the expense of the change in the forms of inclusion close to their melting temperatures under the effect of the sign-changing elastic stresses. The temperature of the peaks provides for making assumptions about the composition of the inclusions.

In the course of the work undertaken, the internal friction of the investigated samples of SnTe and PbTe obtained by the iodide method was studied at the frequencies of the infrasound range. The samples were prepared as plates of a parallelepiped form with dimensions of the order of 15 × 5 × 1 mm³. The results of the measurements are shown in Figures 1 and 2. The first Figure shows the temperature dependences of internal friction of sample SnTe obtained by the iodide method which has passed through the following technological conditions: Sn, Te, and I, of 3 N purity, were used for its preparation. The mass of the elements fed was calculated from the ratio of SnI_2:SnTe = 1:2. Synthesis was carried out at 450 °C for 8 hours by the use of vibration mixing. Isothermal annealing of the material obtained was carried out in the course of 30 hours in a hydrogen flow at 400 °C. The hydrogen was subjected to additional purification by means of a palladium diffuser. In order to alleviate the process of treatment of the materials obtained, the annealing mass was periodically

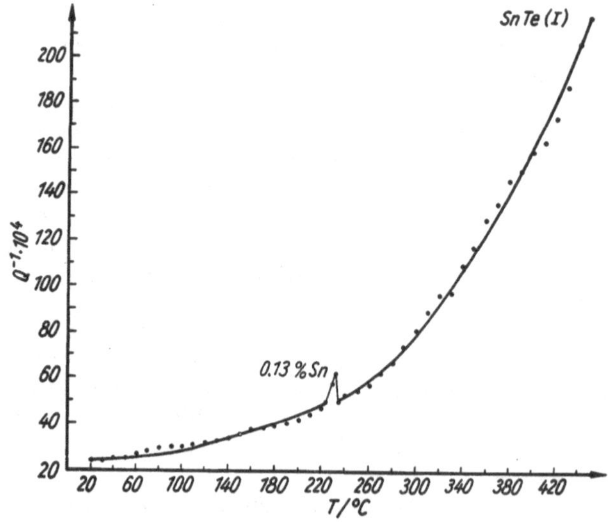

Fig. 1. Temperature dependences of the acoustic attenuation (internal friction) for samples of SnTe(I). At the meltig point of pure tin (232 °C) there is a characteristic peak to which correspond 0.13 v.p.c. tin in excess

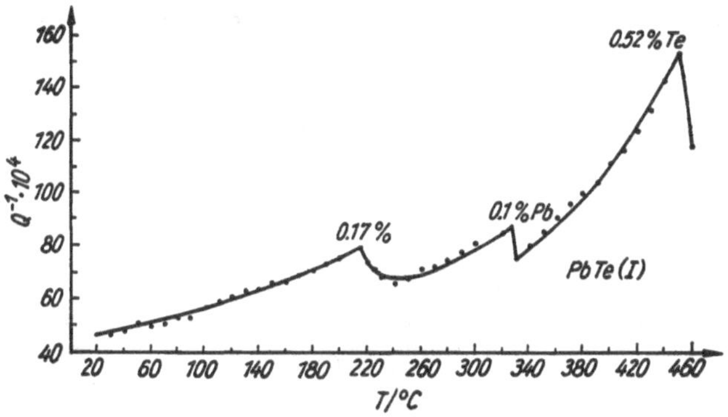

Fig. 2. Temperature dependences of the acoustic attenuation (internal friction) for samples of PbTe(I). There are 3 characteristic peaks at the melting points of lead (327 °C), of tellurium (450 °C), and of an unidentified substance (215°C) with content of these substances in excess 0.1 v.p.c., 0.52 v.p.c., and 0.17 v.p.c., respectively

taken every 10 hours from the reactor and was crushed in an agate mortar. At the final stage of purification of the reaction product from more volatile compounds (tin iodides) annealing takes place at $T = 450$ °C, under conditions of dynamic vacuum (1×10^{-5} Torr) for 3 hours. The formation of the compact ingot, from which the sample tested was cut, was done by briefly melting the SnTe powder at 830 °C for 10–15 minutes. The values of the Hall concentration and the mobility of the carriers and of the charge in the sample after treatment amounted to $P_{300\,K} = 5 \times 10^{19}$ cm^{-3} and $\mu_{300\,K} = 850$ cm^2/V.S. These values are close to the data corresponding to the metallic boundary of the homogeneity area, which is normally reached by continuous saturating annealings in vapours of the alloying composition with excess content of the metallic component.

Figure 1 shows that close to the values of the melting point of pure tin ($T = 232$ °C) of the dependence of the internal friction on the temperature there is a characteristic peak, corresponding to an insignificant content of excess tin (0.13 v.p.c. according to the evaluation given by MITROKHIN et al.).

Analogous investigations with PbTe samples are shown in Figure 2. Here the technological operations involved in obtaining lead telluride were similar to the conditions under which SnTe was obtained. PbI_2 was used as solvent, and the charge consisted of pure elements in the same proportion. The temperature of preliminary synthesis was 600 °C, while purification in a flow of hydrogen in dynamic vacuum was carried out at 550 °C. The final formation of the ingot was made by Bridgeman's method. The electrophysical parameters at 300 K were: $n_{300\,K} = 5 \times 10^{19}$ cm^{-3} and $\mu_{300\,K} = 900$ cm^2/V.S.

It can be seen from the temperature dependences of the acoustic attenuation (Fig. 2) that there are three characteristic peaks in PbTe, two of them coinciding with the melting points of lead ($T = 327$ °C) and tellurium ($T = 450$ °C). The concentration of inclusions, calculated by the method of MITROKHIN et al., is 0.1 v.p.c. Pb and 0.52 v.p.c. Te.

It must be pointed out that in Figures 1 and 2 there are no peaks connected with the melting point of iodine ($T = 113$ °C) and of certain iodides — SnI_2 ($T = 320$ °C), SnI_4 ($T = 140 - 145$ °C) and PbI_2 ($T = 386 - 412$ °C). This demonstrates sufficiently effective purification of the materials obtained from the indicated iodides in the process of thermal treatment.

At the same time, the temperature dependence of the acoustic attenuation in the PbTe sample has a peak at 215 °C, whose nature we have been unable to determine definitely for the time being.

It is obvious from the results of the measurements that in this particular case there was no complete reaction between lead and tellurium in the PbTe obtained by the iodide method, at the same time when in SnTe there was an excess of only one of the components — Sn. The result on the observed excess of tin is in agreement with and supplements results of research carried out (ASSENOV et al.) to the effect that the separated phase in the flow of the inert gas corresponds basically to the formula SnI_4. This must lead to the formation of excess Sn in the material processed and toward its self-alloying in the process of heat treatment, according to the reaction:

$$2\,SnI_2 \rightarrow Sn + SnI_4\,.$$

In the case of PbTe it is necessary to note the predominant amount of tellurium (0.52 v.p.c.) compared to lead (0.1 v.p.c.). Analogous investigations for this system have been carried out by ASSENOV et al. It was established that the phase separated in the process corresponds to formula PbI_2, i.e. there should be no accumulation of lead in the material at the expense of the separation. In this manner, the difference between the parts of separated lead and tellurium (0.4 v.p.c.) must correlate with the predominantly dissolved lead in the lattice of the PbTe obtained by the iodide method. This process is in agreement with the model examined by ASSENOV et al.

References

ASSENOV, R., MOSHNIKOV, V. A., YASKOV, D. A.: phys. stat. sol. (a) 88, K27 (1985)

ASSENOV, R., MOSHNIKOV, V. A., SAUNIN, I. V., YASKOV, D. A.: Cryst. Res. Technol. 21 (12) 1549 (1986)

ASSENOV, R., MOSHNIKOV, V. A. SAUNIN, I. V., YASKOV, D. A.: Crystal. Res. Technol. 22, 1153 (1987)

GLUSHKOV, E. A., IZMAILOV, N. V., LITVIN, A. A., REMBEZA, S. I., TUZOVSKI, A. M., TURKOV, C. K., YAROSLAVTSEV, N. P.: O mehanizmah obrazovaniya komponenta A^3 v poluprivodnikakh A^3B^5, Academy of Science of the USSR-Reports, seriya Neorganicheskie materiali, N12, (1985)

IZMAILOV, N. V., KRASNOV, V. A., KUZNETSOV, A. S., YAROSLAVTSEV, N. P.: "Mikrovklyucheniya rastvoritelya v epitaksialnikh sloyah kremniya", serija 6, Materiali, N10, Moscow 1985, p. 61—62

KWESTROO, W.: Poluchenie halkogenodov i pniktidov pri niskoi temperature, in: Proceedings: Preparativnie metodi vkhimii tverdogo tela, Moskva (1976), p. 592

MITROHIN, V. I., YAROSLAVTSEV, N. P., IZMAILOV, N. B., REMBEZA, S. I., LISOVENKO, V. D.: USSR-Authors Certificate, N 1179183 (12. 12. 1983)

MOLDOVANOVA, M., TRIFONOVA, E. P., ASSENOV, R., KARAGIOZOV, L.: phys. stat. sol. (a) 58, K47 (1980)

(Received February 12, 1987)

Authors' addresses:

R. ASSENOV
Semiconductor Physics and Technology,
Research Laboratory, University of Sofia
Boulvd. A. Ivanov, No 5
Sofia 1126, Bulgaria

N. V. IZMAILOV, V. A. MOSHNIKOV, N. P. YAROSLAVTSEV, D. A. YASKOV
Leningrad Electrotechnical Institute
Leningrad, USSR

Cryst. Res. Technol.	22	1987	9	1193–1203

Hanspeter HEEGN

Forschungsinstitut für Aufbereitung der Akademie der Wissenschaften der DDR, Freiberg

Model Describing the Resistance Against Structural Changes and the Hardness of Crystalline Solids

On the basis of knowledge concerning the thermodynamic state and the structure of unelastically deformed solids the present paper derives a model for the description of the resistance of solids against mechanically induced structural changes. The concentration of dislocations, the external specific surface, the particle size of primary crystallites and the amorphization are important for characterizing the solids in the state of stress. The results of model calculations are compared with data available from literature on MOHS hardness and microhardness of crystalline solids. The results show a sufficient correlation independent of the nature of solids.

Auf der Grundlage von Erkenntnissen zum thermodynamischen Zustand und zur Struktur unelastisch deformierter Festkörper wird in der vorliegenden Arbeit ein Modell zur Beschreibung des Widerstandes von Festkörpern gegen mechanisch induzierte Strukturveränderungen abgeleitet. Die Versetzungskonzentration, die äußere spezifische Oberfläche, die Teilchengröße der Primärkristallite und die Amorphisierung sind für die Charakterisierung des Festkörpers im Beanspruchungszustand von Bedeutung. Die Ergebnisse der Modellrechnungen werden mit den aus der Literatur zugänglichen Angaben zur MOHS-Härte und zur Mikrohärte kristalliner Feststoffe verglichen. Die Ergebnisse zeigen eine befriedigende Korrelation unabhängig von der Natur der Stoffe.

1. Introduction

In recent years the opinions on the fundamental processes in the action of mechanical forces on crystalline solids became increasingly founded, since besides the classical theory of elasticity much attention has been paid to the knowledge of solid-state physics and chemistry and especially of fracture physics. However, a closed theory involving elastic and inelastic processes could not yet be spoken of. The physics of the subprocesses occurring at an atomic level allows only a qualitative description up to now, since the theoretically founded continuum mechanics cannot be applied. Under stress, deformation and fracture defects are caused in the solid, the nature and concentration of which being a function of material properties at the corresponding temperature and stress conditions. There are complicated relationships, the measuring-technological registration of which causes some problems. This state is also represented by the great variety of conceptions and testing procedures for characterizing the mechanical properties. The conceptions strength, hardness, brittleness, plasticity, elasticity and cleavage are used without being defined clearly and sufficiently generally. On the basis of the knowledge concerning the thermodynamic state and structure of inelastically deformed solids the present paper derives a model for the description of the resistance of solids against mechanically induced structural changes. The results of model calculations are compared with data available from literature on MOHS hardness and micro-hardness of crystalline solids.

2. Derivation of the model

For the calculation of the stresses in a solid under mechanical force there are numerous theories demonstrating the complicated relationship between stress geometry and distribution of compaction, tensile and shear stresses in a stressed volume. This paper is intended to give a qualitative evaluation of relationships.

According to the laws of classical solid-state mechanics the energy concentration in a stressed volume within the elastic range is proportional to the square of the acting forces (Hooke's law):

$$U_E = \frac{p_E^2}{2E} \quad \text{and} \quad U_K = \frac{p_K^2}{2K} \quad \text{and} \quad U_G = \frac{p_G^2}{2G}, \tag{1}$$

respectively.

U_E, U_K, U_G energy concentration in the volume
p_E, p_K, p_G deformation force under corresponding tension, compression or shear stress
E, K, G Young's modulus, compression modulus and shear modulus, resp.

If the force impact exceeds the elastical range, accumulation of defects occurs in the solid, the nature and concentration of which being dependent on specific material properties and deformation quantities. In the stressed state a certain defect structure occurs the easier, the lower the specific energy of defects. Thus the concentration of defects with small specific energy is greater than those with high specific energy. The energy concentration in the stressed volume is the higher, the greater the specific energies of defects, where the absolute value of this parameter is inversely proportional to the sum of reciprocal values of the energies of the single defects (HEEGN, ILGEN).

$$\sum_{i=1}^{m} \frac{1}{E_i} = \frac{1}{E_\Sigma}. \tag{2}$$

The elastic range of deformation of a solid is exceeded, if at least in a microvolume the energy concentration E_Σ is reached. For this limiting case

$$E_\Sigma = \frac{p_E^{*2}}{2E} = \frac{p_K^{*2}}{2K} = \frac{p_G^{*2}}{2G} \tag{3}$$

holds true.

Thus this critical energy concentration can be defined as resistance against structural changes. For the critical stresses p^*

$$p_E^* = \sqrt{2EE_\Sigma} \quad \text{and} \quad p_K^* = \sqrt{2KE_\Sigma} \quad \text{and} \quad p_G^* = \sqrt{2GE_\Sigma} \tag{4}$$

is valid.

By means of the known relationship between elasticity constants:

$$E = 2G(1+\nu) = 3K(1-2\nu) \tag{5}$$

it is clear, that the critical stresses depending on the mode of stress must be different and that the Poisson's ratio ν is much important in the evaluation of the effect of different stress in a certain material system.

3. Calculation of the energies of defects

In real systems there is the problem to differently determine the defects produced under stress, deformation and fracture as regards their mode and concentration.

Here the following, probably decisive and also measuring-technically provable defect structures are discussed (HEEGN):

- the concentration at dislocations c_V with the energy E_V
- the external specific surface c_A with the energy E_A
- the particle size of primary crystallites c_K with the energy E_K
- the amorphization or phase modification c_p with the energy E_p.

Further defects and especially short-time activated states are also important, but can only be quantitized theoretically and experimentally with difficulties. The presented approximation is not intended to further observe these defects.

Dislocations

The energy of dislocations is given by the work required for their production in the undisturbed isotropic medium. For dislocation energy per unit of dislocation length (KLEBER et al.)

$$U_S = \frac{Gb^2}{4\pi} \ln \frac{R}{r_0} \quad \text{and} \quad U_{St} = \frac{Gb^2}{4\pi(1-\nu)} \ln \frac{R}{r_0} \tag{6}$$

is obtained.

U_S energy of screw dislocations
U_{St} energy of edge dislocations
G shear modulus, ν Poisson's ratio
b contribution of Burgers vector, comparable with atomic distance
r_0 radius of dislocation nucleus, comparable with atomic distance
R crystallite dimension, primary particle size.

The evaluation of the rate of the specific dislocation energy related to volume is based on the fact that by dislocation not only the lattice components are concerned in the dislocation line, but also at least the closest neighbours. Thus the coordination number of the lattice must be considered. The volume-related dislocation energy is given by

$$E_{V,S} = \frac{G}{4\pi KZ} \ln \frac{R}{r_0} \quad \text{and} \quad E_{V,St} = \frac{G}{4\pi KZ(1-\nu)} \ln \frac{R}{r_0} \tag{7}$$

KZ coordination number.

Edge and screw dislocation differ only by the factor $1/1-\nu$, where usually both dislocation types always occur in the mixture, such that this factor as a mean value is only slightly greater than 1. For the ratio between crystallite dimension and dislocation distance, on the average, a value of $R/r_0 = 50$ and $\ln R/r_0 = 4$, resp., may be determined.

On the basis of the compression modulus summarized in tables for many materials and of the relationship between moduli of elasticity given by eq. (5) the dislocation energy can be approximately calculated by the following relation:

$$E_V = 0.5 \frac{K(1-2\nu)}{KZ(1+\nu)} \tag{8}$$

Outer specific surface

The disorder energy of the lattice components being situated at the boundary to the surrounding medium corresponds to the surface energy of solids. By means of the following relation the relationship between the energy related to the surface and to volume is given:

$$E_A = \sigma_A N_L^{1/3} V_M^{-1/3} . \tag{9}$$

σ_A specific surface energy, V_M molar volume
N_L LOSCHMIDT number.

The determination of the specific surface energies at present is still a problem. For chemical compounds with simple lattices there are calculation results, however they differ very much from author to author depending on the respective assumptions made. For the most part of solids only rough estimations are known. The anisotropy of the solids additionally causes a difference between the surface energies of crystalline faces according to their position in the lattice. In order to calculate the specific surface energies of solids the present paper describes a method starting from mechanical material data. The model is based on the assumption, which has also been proved by DZAPARIDZE, that the action of the free surface forces leads to a volume contraction, the extent of which is determined by the present respective valence electrons. According to the surface coverage by lattice components thus the relative volume contraction can be determined, and by means of the compression modulus available from literature the specific surface energy can be calculated. The present case is concerned only with the estimation of medium values of surface energies.

This is given by

$$\sigma_A = \frac{1}{2} K \left(\frac{\Delta V}{V_0}\right)^2 V_M^{1/3} N_L^{-1/3} . \tag{10}$$

The relative volume contraction is given by

$$\frac{\Delta V}{V_0} = \frac{r_0 - r_R}{2 r_0} . \tag{11}$$

r_0 atomic and anion radius, resp., r_R = radius of the inner atomic trunk.

The calculation of surface energies by means of the presented model results in a correct order both by tendency and by the absolute value in comparison with references from literature. Data on selected materials of the most different compound types and elements are summarized in Table 1.

Particle size of primary crystallites and amorphisation

The disorder energy of lattice components being situated at the grain boundaries of primary crystallites can rather probably be compared with the melting heat related to the lattice components contained in the boundary (MEYER), since a quasi-amorphous structure can be assumed. An amorphisation is frequently proved especially by X-ray methods, if crystalline solids are subjected to high mechanical stress. This X-ray amorphisation is caused by small primary particles, which are not diffractable for X-radiation. In many cases a direct proportional relationship between X-ray amorphisation and primary particle size could be proved (HEEGN et al.) such that it is easy to assume that both effects cannot be separated. Thus $E_K = E_p$ is valid, where the melting enthalpy is considered to be a suitable characteristic for the amorphous state.

4. Relative contribution of defects

Since according to eq. (2) the relative concentration of defects in the stressed state is inversely proportional to the specific defect energy, an evaluation of the different substances can be carried out by surface energy, dislocation energy and melting

Table 1
Surface energies of selected solids and preliminary data for their calculation

material	K	r_0	r_R	$\Delta V/V_0$	σ_A erg/cm²				
	kN/mm²	Å			th.	ref.			
								(100)	(110)
LiF	68.2	1.36	0.72	0.235	478	417		142	568
NaF	49.3	1.36	0.72	0.235	396	275	306	216	555
NaCl	24.4	1.81	0.99	0.227	223	152	266	158	354
NaBr	20.5	1.95	1.14	0.208	166	128	177	138	304
KCl	18.5	1.81	0.99	0.227	189	110		141	298
BeO	220	1.40	0.74	0.236	1469	2250			
MgO	158	1.40	0.74	0.236	1170	1459		360	924
CaO	102	1.40	0.74	0.236	862	979		509	879
ZnO	133	1.40	0.74	0.236	1067				
PbS	52	1.84	1.04	0.217	458	520			
$SiO_2$¹)	200*)	1.40	0.74	0.236	1863	2350			
ZrO_2	180	1.40	0.74	0.236	1648	2250			
$TiO_2$²)	179	1.40	0.74	0.236	1569				
Fe_3O_4	182	1.40	0.74	0.236	2137	1800			
α-Fe_2O_3	172	1.40	0.74	0.236	1769	1650			
α-Al_2O_3	322	1.40	0.74	0.236	3127	2300			
$MgCO_3$	100	1.40	0.74	0.236	998				
$CaCO_3$³)	74	1.40	0.74	0.236	813	950			
$BaSO_4$	58	1.40	0.74	0.236	713	800		(111)	(110)
CaF_2	83.6	1.36	0.72	0.235	795	1050		1082	543
α-Fe	175.8	1.26	0.64	0.246	1211	3495	2758		
Ni	194.6	1.24	0.72	0.210	954	3480	2348		
ε-Co	190.2	1.25	0.74	0.204	880	2725			
Cu	142.2	1.28	0.72	0.219	776	2725	1629		
Pb	42.1	1.70	0.84	0.253	420	830	538		
Sn	54.5	1.62	0.71	0.281	646	1200			
Zn	60.7	1.34	0.74	0.224	376	900			
Al	75.0	1.43	0.50	0.325	1010	1800	996		
C⁴)	637.3	0.77	0.15	0.403	9227	7735			
Si	104	1.38	0.41	0.351	1740	2100			
ref.	LANDOLT	PAULING			HEEGN		BENSON	SPITZER	

*) estimated value, because a reliable reference is not available.
¹) quartz, ²) rutile, ³) calcite, ⁴) diamond.

energy. The energy values of selected substances are given in Table 2. The relative concentration of defects of selected substances is presented in Figure 1 by a triangular diagram. The absolute numerical values are not the essential ones but the tendencies. Accordingly the metals e.g. form only hardly new surfaces, they rather form dislocations under stress and are transferred into the amorphous state by primary particle reduction. This trend is most pronounced in the case of tin, lead and aluminium. Alkali halides and simple oxides, however, form easier new surfaces and dislocations, where the amorphisation is reduced. In the case of the other investigated oxides, diamond, silicon as well as carbonates and sulfates a medium position is shown, since the single defect energies do not differ much from each other. It is interesting to

Table 2

Energies of defect structures, resistance against structural changes and critical stresses

material	KZ	ν	E^s_V	E^s_P kJ/cm³	E^s_A	E_Σ	P_K	P_E kN/mm²	P_G
LiF	6	0,29	1,85	2.69	1.88	0.69	9.7	10.9	6.8
NaF	6	0.20	2.05	2.25	1.36	0.60	7.7	10.3	4.6
NaCl	6	0.21	0.98	1.07	0.63	0.28	3.7	3.4	2.6
NaBr	6	0.21	0.82	0.72	0.44	0.20	2.9	3.8	2.4
KCl	6	0.14	0.97	0.68	0.48	0.22	2.9	4.2	2.8
BeO	4	0.26	10.48	8.57	6.13	2.66	34.2	41.1	25.9
MgO	6	0.24	5.53	6.87	4.40	1.81	23.9	29.9	18.9
CaO	6	0.23	3.73	3.04	2.84	1.05	14.6	18.6	11.9
ZnO	4	0.34	3.97	3.63	3.70	1.25	18.2	17.9	11.0
PbS	6	0.19	2.26	0.55	1.22	0.32	5.8	7.9	5.1
SiO_2[1])	4	0.14	15.86	3.06	5.56	1.75	26.5	38.9	25.7
ZrO_2	8	0.2	5.62	4.06	5.01	1.60	24.0	32.2	20.8
TiO_2[2])	6	0.2	7.46	3.45	4.98	1.60	23.9	32.0	20.7
Fe_3O_4	4	0.27	8.24	3.06	5.07	1.55	23.8	27.9	17.5
α-Fe_2O_3	6	0.19	7.46	2.96	4.79	1.47	22.5	30.7	19.8
α-Al_2O_3	6	0.24	11.25	4.27	8.97	2.30	38.5	48.1	30.6
$MgCO_3$	6	0.25	3.33	2.16	2.78	0.89	13.3	16.3	10.3
$CaCO_3$[3])	6	0.26	2.35	1.35	2.06	0.61	9.5	11.4	7.2
$BaSO_4$	9	0.25	1.29	0.79	1.62	0.38	6.6	8.1	5.2
CaF_2	8	0.22	2.40	1.21	2.31	0.60	10.0	13.0	8.3
α-Fe	8	0.29	3.57	2.18	5.32	1.08	19.5	21.9	13.6
Ni	12	0.30	2.49	2.70	4.30	1.00	19.7	21.6	13.4
ε-Co	12	0.31	0.30	2.31	3.96	0.89	18.4	19.7	12.2
Cu	12	0.34	1.51	1.96	3.65	0.69	14.0	13.7	8.4
Pb	12	0.43	0.35	0.26	1.34	0.13	3.3	2.1	1.3
Sn	6	0.33	1.16	0.43	2.15	0.27	5.4	5.5	3.4
Zn	12	0.24	1.06	0.80	1.52	0.35	6.5	8.2	5.2
Al	12	0.34	0.75	1.07	3.96	0.40	7.7	8.7	4.6
C[4])	4	0.10	57.94	40.44	51.76	16.31	144	223	151
Si	4	0.28	4.47	3.85	6.41	1.56	18.5	20.7	13.0

ref. LANDOLT D'ANS-LAX

[1]) quartz, [2]) rutile, [3]) calcite, [4]) diamond.

note that quartz shows a strong amorphisation tendency and dislocation formation is strongly reduced. This could be confirmed by X-ray methods applied to the investigation of intensely mechanically stressed quartz, in that lattice distortions could not be observed by STEINIKE. The categorisation of the different substances corresponds to experience. The presented model, however, allows quantitative calculations.

5. Comparison with Mohs hardness

In order to test the performance of the presented model a comparison with the characteristics of strength and hardness of materials referred to in literature is recommended. In the manuals of mineralogy f.e. ROESLER hardness is generally understood as the value of resistance, which a mineral withstands an outer mechanical stress,

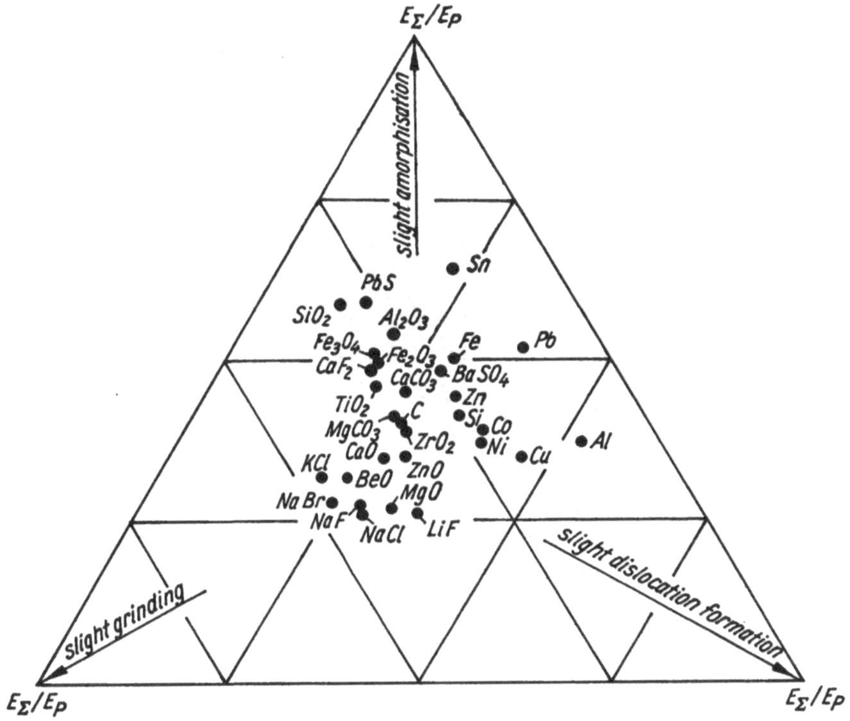

Fig. 1. Relative contribution of defect structures under mechanical stress

especially the scratching or the indentation of another solid body. In structural regard hardness is naturally dependent on lattice structure, especially on bond type and on atomic and ion radii, resp., of the respective crystal type. Despite century-old use the term hardness is not yet unambiguously defined at present. Different methods of scratch, drilling, pressure and abrasive hardness represent different physical properties (TERTSCH) and therefore are not comparable by their numerical values. However it is remarkable that the relatively rough and essentially qualitatively handled method of abrasion testing according to Mohs is most useful for diagnostic purposes. Nevertheless there were numerous trials to stronger quantify this method and on the basis of theoretical considerations to develop the relationship with the structure of materials (ENGELHARDT et al.).

In this connection it is useful to mention the semi-empirical method by POVARENNYCH for the calculation of hardness from the effective valencies of anions and cations, from the atomic distances as well as coefficients for coordination numbers, for covalent bond strengths, for different valency types and for certain element groups. By means of this method a remarkably well agreement with experimental values has been reached. Using these Mohs hardness values introduced by POVARENNYCH the values theoretically calculated in the present paper for the resistance against mechanically induced structural changes are to be compared.

Figure 2 shows the respective relationship. Independently of the nature of materials a rather good fit to the curve drawn in the scheme could be stated. As expected, there is no linear relation, since the Mohs hardness scale ought to be essentially extended above all with harder substances. This is also underlined by the extremely high value of $E_\Sigma = 16.3$ kJ/cm³ for diamond. This value has not been taken in the presentation

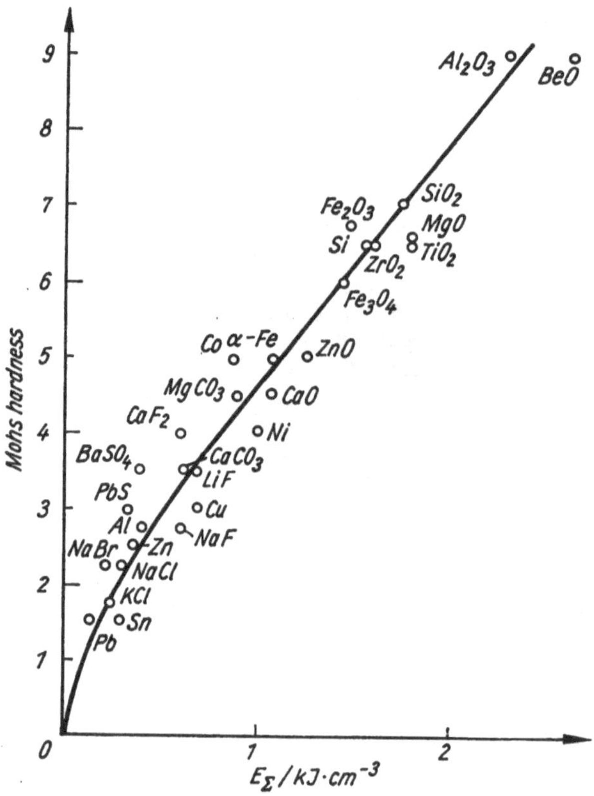

Fig. 2. Relationship between Mohs hardness and the theoretically calculated resistance against structural changes

of Figure 2, for especially between the Mohs hardness 9 and 10 the real difference is much greater than expressable by a hardness degree. In order to match this situation some authors (CHRUSCOV) propose a hardness scale exceeding 10 and reaching 15.

6. Comparison with microhardness

Besides Mohs hardness the so-called microhardness has been introduced to characterize the mechanical properties of crystals and materials, where the pyramid method by Vickers has been generally applied. In this method quantified results are obtained by indenting a diamond pyramid at an opening angle of 136° and microscopic measurement of the indentation. Only very small regions of the solid body are stressed, such that macroscopic structural defects do not become effective, but only the crystal-typical properties. In equilibrium after the indentation of the diamond pyramid the resistance against structural changes and deformations just equalizes the force acting on the diamond. Between the deformed volume of the solid and the acting force there is a proportionality depending on microhardness. In the evaluation of microhardness the anisotropy of the surfaces of a solid body must be considered. Polycrystalline materials without preferred orientation of single crystals can be characterized only by one hardness value as isotropic solids. However, it must be mentioned, that despite

Table 3
Variations of the microhardness data (POVARENNYCH)

material	Mohs hardness	microhardness kJ cm^{-3} or $kNmm^{-2}$	
		range of variation	probable value
talc (001)	1	0.024 ... 0.11	0.02 ... 0.03
gypsum	2	0.36 ... 0.90	0.5 ... 0.6
calcite	3	1.05 ... 1.72	1.36 ... 1.45
fluorite (111)	4	1.64 ... 2.60	1.7 ... 2.0
apatite	5	5.36 ... 6.90	5.0 ... 6.0
orthoclase	6	7.40 ... 8.50	7.6 ... 8.0
quartz (1011)	7	10.0 ... 14.6	10 ... 11
topaz (001)	8	14.27 ... 18.00	14 ... 16.5
corundum (1120)	9	20.50 ... 22.00	20.5 ... 22
diamond	10(15)	100.6	100

the different hardness to be expected of different crystal faces of anisotropic crystals the experimentally determinable microhardness anisotropies are only weakly pronounced. Many crystals therefore can be characterized with good approximation by only one hardness value. This quasi-isotropic behaviour is caused by the fact, that in the close zone beneath the surface of the indentation of the diamond pyramid the crystalline order is nearly destroyed, and the active solid states discussed in the preceding sections occur. Thus a good agreement between the theoretically calculable resistance against structural changes being mechanically induced and microhardness can be expected. However, the microhardness values discussed in literature show a wide range of variation.

Table 3 presents as an example variations of microhardness, probable values of microhardness and Mohs hardness available from POVARENNYCH for the reference minerals of Mohs's scale.

The relationship between microhardness and the theoretically calculated resistance against structural changes is shown in Figure 3. There is a good fit into the curve, such that actually with the value of the resistance against structural changes the material behaviour under mechanical stress in the microrange can be estimated theoretically from known material data.

7. Concluding remarks

Based on the specific defect energies of the defects preferredly expected under mechanical stress in the inelastic range a model for the calculation of the resistance of solids against mechanically induced structural changes and for the estimation of relative defect concentrations in the stressed state is derived. For selected solids the data of defect energies of surface, of dislocations and amorphisation are given and discussed. The categorisation of the different materials corresponds to experience. By means of the theoretically derived values of the resistance against structural changes the critical stresses for the transition of elastic deformation into inelastic structural changes can be calculated in dependence on stress type.

The comparison with the values available in literature for scratch hardness according to Mohs and microhardness results in a sufficient correlation independent of the nature of materials.

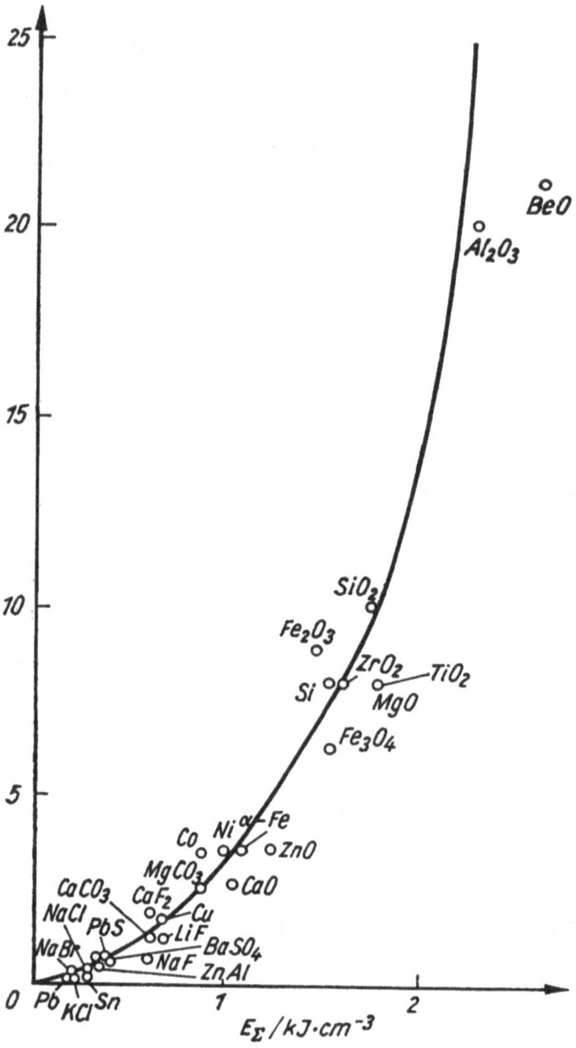

Fig. 3. Relationship between microhardness and the theoretically calculated resistance against structural changes

References

BENSON, G. C., YAN, K. S.: The Solid-Gas-Interface, New York 1967, Vol. 1, pp. 203–269

CHRUSCOV, M. M.: Zavod. Labor. **15**, 213 (1949)

D'ANS, LAX, E.: Taschenbuch für Chemiker und Physiker, Berlin—Heidelberg—New York 1967, Vol. 1

DZAPARIDZE, P. N.: Banicke listy, **3**, 17 (1975)

ENGELHARDT, W. O., HAUSSUHL, S.: Fortschr. Mineral. **42**, 5 (1965)

HEEGN, H.: Banicke listy, specialissue 1976, p. 58

HEEGN, H., BERNHARDT, C., GOTTSCHALK, J., HUSEMANN, K.: Chem. Techn. **26**, 696 (1974)

HEEGN, H., ILGEN, S.: Freiberger Forschungshefte A 700, pp. 248–264 (1985)

Kleber, W., Meyer, K., Schoenborn, W.: Einführung in die Kristallphysik, Berlin 1968, pp 180–183
Landolt-Börnstein: Mechanisch-thermische Zustandsgrößen, Berlin, Göttingen, Heidelberg, New York. Zweiter Band 1. Teil (1971), Neue Serie III Bd. 1 (1966), III. Bd. 1 (1969)
Meyer, K.: Physikalisch-chemische Kristallographie, Leipzig 1977
Pauling, L.: Die Natur der chemischen Bindung, Weinheim/Bergstr. 1962
Povarennych, A. S.: Die Härte der Minerale, Kiev 1963 (Russ.)
Rösler, H. J.: Lehrbuch der Mineralogie, Leipzig 1979
Spitzer, D. P.: J. phys. chemistry **66**, 8 (1962)
Steinike, U., Hennig, H.-P., Richter-Mendau, I., Kretzschmar, U.: Cryst. Res. Technol. **17**, 1585 (1982)
Tertsch, H.: Die Festigkeitserscheinungen der Kristalle, Wien 1949

(Received February 23, 1987)

Author's address:

Dr. sc. Hanspeter Heegn
Forschungsinstitut für Aufbereitung
Akademie der Wissenschaften der DDR
Straße des Friedens 40
Freiberg
DDR-9200

Book Review

J. Krautkrämer, H. Krautkrämer

Werkstoffprüfung mit Ultraschall

Springer-Verlag, Berlin, Heidelberg, New York, London, Paris, Tokyo, 1986, 5. völlig überarbeitete Auflage, 708 Seiten, 564 Abbildungen, 10 Tafeln, Gebunden DM 298. –, ISBN 3-540-15754-9

Das Buch ist in 4 Teile übersichtlich gegliedert. Im Teil A werden die physikalischen Grundlagen der Ultraschallwerkstoffprüfung dargestellt, was nahezu 30% des Gesamtumfanges ausmacht. Dabei verzichten die Autoren auf ein hohes theoretisches Niveau und vermitteln anwendungsbereites Wissen. Gegenüber den vorangegangenen Auflagen werden besonders die Grenzflächenreaktionen bei Reflexion und Transmission, das reale Schallfeld und seine Transformation bei Reflexionserscheinungen sowie die Verfahren des Ultraschall-Senden und -Empfangens auf dem neuesten Erkenntnisstand beschreiben.

Der Teil B führt den potentiellen Anwender anschaulich und mit meßtechnischem Einfühlungsvermögen in die geräte- und verfahrenstechnischen Grundlagen der Ultraschallprüftechnik ein und gibt auch dem anspruchsvollen Wissenschaftler neue Anregungen zur Weiterentwicklung, was besonders in der Beschreibung der Bild- und Rekonstruktionsverfahren zum Ausdruck kommt. Die Fragen der Automatisierung und des Computereinsatzes bleiben dabei weitgehend noch im Hintergrund. Der Teil C enthält eine zusammenfassende Darstellung wichtiger prüfpraktischer Informationen, die sich auf die Ankopplung, die Reflexionsanalyse und den Ungänzennachweis beziehen. Bezüglich der Mehrfrequenzanalyse und der Spektroskopie wird auf Informationsreserven verwiesen. Die Betrachtung wird mit Hinweisen zur Organisation der Prüfung und zur Aus- und Weiterbildung abgeschlossen, wobei die internationale Situation ausgewertet wird. Der Teil D ist einer Vielzahl von spezifischen Prüfaufgaben gewidmet, was über 30% des Gesamtumfanges ausmacht und den Charakter des Buches wesentlich prägt. Hier findet der Anwender die grundlegenden prüftechnologischen Hinweise mit einer Zusammenstellung der internationalen Normen. Schwerpunktmäßig werden dabei neben Guß- und Schmiedeteilen sowie Schweißnähten die Prüfungen von KKW-Komponenten behandelt. Aber auch viele Sonderprüfaufgaben werden mit hohem Informationsgehalt dargestellt.

Insgesamt kann eingeschätzt werden, daß das Buch eine gelungene Symbiose zwischen anwendungsbereiten Erkenntnissen und Vorlaufwissen ausweist. Die 1752 Literaturstellen und die über 100 internationalen Normen machen das Buch zu einem wertvollen Arbeitsmittel für den Prüfingenieur und geben zugleich zahlreiche Anregungen für die wissenschaftliche Weiterentwicklung der Prüftechnik. Das Buch hat gegenüber den vorangegangenen Auflagen eine neue Qualität erreicht und ist aus der Forschung, Lehre und Prüfpraxis nicht wegzudenken. Die vorwiegend auf die Prüfpraxis ausgerichtete Grundkonzeption gibt dem Buch trotz des permanenten Fortschrittes der Computertechnik und Mikroelektronik einen Langzeitwert.

U.-D. Hünicke

Cryst. Res. Technol.	22	1987	9	1205–1209

A. BALARAMAIAH, K. J. PRATAP, V. HARI BABU

Department of Physics, Osmania University, Hyderabad, India

Microhardness Studies of $K_{1-x}Rb_xCl$ Crystals

The microhardness measurements have been made on pure crystals of KCl, RbCl and mixed crystals of $K_{1-x}Rb_xCl$. Microhardness varies non-linearly with composition being maximum near at the eqimolar composition. Microhardness values calculated from lattice parameter data are in good agreement with experimental data. The difference in the size of the ion constituting the mixed system is responsible for the internal strains giving rise to imperfections and inturn responsible for the non-linear variation of microhardness with composition.

Introduction

Microhardness is a test for characterising the mechanical behaviour of minerals and materials. The divalent impurities present in alkali halides strongly effect the mechanical properties provided the impurity does not form a visible precipitate. The micro structural and hardness studies on KCl crystals doped with alkali earth impurities (LAKSHMIPATHI RAO, HARI BABU) have shown that dislocations act as nucleation sites for the formation of visible precipitates effecting the hardness of the crystal. The effect of impurity on the flow stress of LiF crystals have been studied (JOHNSTON) and found that divalent impurity ions are more effective than monovalent impurities. Ammonium halides doped with copper resulted (PRATAP, HARI BABU) in enormous increase of hardness. In alkali halide mixed crystals of KCl–KBr and KBr–KI (SUBBA RAO, HARI BABU) local strains develop due to the difference in the ionic size and effect the microhardness of the crystals. No results have been reported so far on the microhardness of KCl–RbCl mixed system which is forming a complete solid solution. An attempt is therefore made to study the microhardness of these mixed crystals and the present paper out lines:

i) The variation of microhardness with Load in each composition of mixed crystal.
ii) Variation of microhardness with composition.
iii) Calculation of microhardness values from lattice parameter data and comparision with experimental values.
iv) The role of ionic size on the microhardness.

Experimental

Single crystals of KCl, RbCl and their mixed compositions were grown by Kyropoulos and slow cooling methods from melt. The starting materials used for the growth of these crystals are BDH analar grade salt (KCl) and Fluka granites (RbCl) of high purity. The components after weighing separately grounded into a fine powder and then mixed well The mixture was taken into a platinum crucible and heated to a temperature 60 °C above. the melting point. The crystals were pulled at the rate of 1 cm/hr from the melt. In the slow cooling method a quartz ample of length 4—6 cm and diameter 1 cm ending conically to a point was taken and filled with the mixture. The mixture was heated to a temperature above its melting point and kept for about 5 hours at that temperature to obtain uniform solid solution. Then it was allowed to cool slowly at the rate of 2 °C /hr till it was solidified. After complete solidification it was cooled further at the rate of 5 °Chr. After the growth all the crystals were annealed for 12 hours under identical condition at 600 °C.

The microhardness measurements were performed by Vicker's indentor attached to a universal research microscope (Carl-Zeiss, Jena). The Loadswere varied from 2.5 to 60 p in steps of 2.5 p and 5 p. The time of indentation was maintained as 10 seconds. At least 10 indentations were made with each load on freshly cleaved crystal. The diagonal of the indentations was measured and the average value of such measurements was taken. The microhardness was calculated using the expression

$$H_v = \frac{1.8544\ \text{p}}{d^2}\ \text{kg/mm}^2$$

where p is the applied load and d being the diagonal of indentation.

The lattice parameters of all these compositions were determined using X-ray powder method and the values were used for the calculation of microhardness using the relation (SUBBA RAO, HARI BABU).

$$H = \text{Const}\ Xr^{-m} + KN_aN_b$$

where KN_aN_b denotes the disorder parameter and K is the coefficient of hardness. N_a and N_b are the molar concentration of the components.

Results

Microhardness measurements had been carried out on freshly cleaved samples at room temperature in (100) direction. Figure 1 shows variation of microhardness with oad for KCl, RbCl and varied composition of $K_{1-x}Rb_xCl$ mixed crystals. From the figure it can be observed that the microhardness is higher at lower loads and decrease

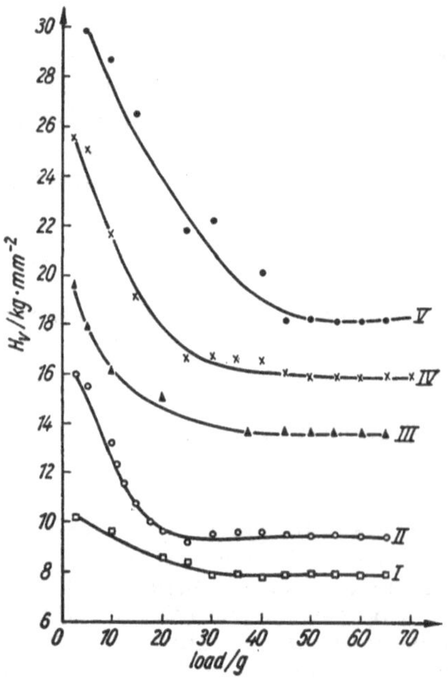

Fig. 1. Variation of microhardness with load for pure and different compositions of KCl and RbCl I – RbCl, II – KCl, III – 12% RbCl, IV – 63% and V – 50% RbCl in KCl

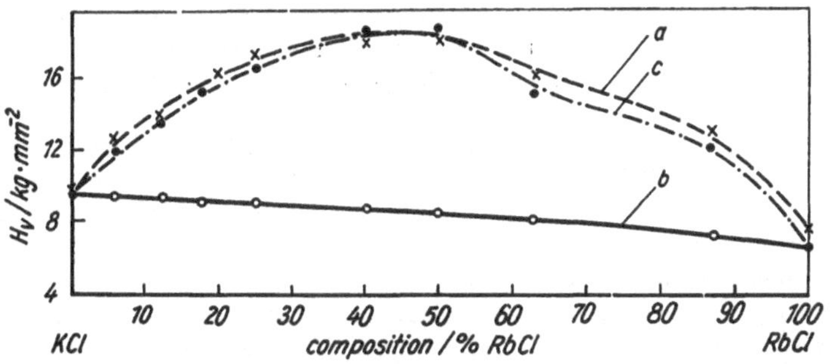

Fig. 2. Variation of microhardness with composition of RbCl in KCl (a) experimental at load 50 p, (b) calculated using the equation 2, (c) calculated using equation (3)

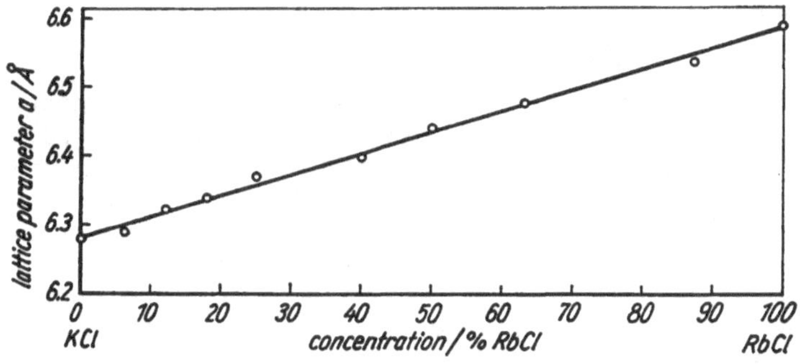

Fig. 3. Variation of lattice parameter with composition of $K_{1-x}Rb_xCl$ crystals

asymptotically to a steady value at higher Loads. From the figure one also notices that steady values are attained at higher loads in mixed crystals when compared to KCl and RbCl. Figure 2 shows the variation of microhardness with composition. It shows that hardness increases with composition and attains a maximum value near equimolar composition. Figure 3 represents the variation of lattice parameters of mixed crystals with composition. The figure shows that the lattice parameter varies linearly with composition and obeys Vegard's rule.

Microhardness is calculated from the lattice parameter data and the calculated results are in good agreement with experimental values as represented by curve C of Figure 2.

Discussion

The generally accepted definition of microhardness is that it is the resistance offered to dislocation motion. The load P is related to the diagonal of indentation d (MAYER et al., 1908) by the relation

$$P = ad^n , \tag{1}$$

where a is constant and n is nearly equal to 2. However, n is usually found less than 2 especially in the low load hardness region. Figure 1 shows the variation of hardness H_v kg/mm^2 with load for both pure crystals of KCl, RbCl and their mixed compositions in which the hardness particularly at the low load region increased as the load is decreased. The variation at very smaller loads (< 5 p) is not so significant because it may arise due to vibrations of the indentor when it touches the surface of the crystal or may be due to the error involved in measuring small impressions (YOST). However, nearly at 5p onwards the hardness factor falls sharply to a steady value and all the curves are asymptotic to the load axis at higher loads. This consistancy of microhardness starts at different values for different compositions and being highest for the equimolar composition of KCl–RbCl. This trend suggests that the surfaces involved are too defective. Further it is observed that the volume of the material involved decreases with decreasing load which implies an increasing contribution of the surface towards the hardness factor (UPIT, VARCHENYA) in case of KCl, KBr and KI crystals.

The resistance offered to dislocation motion is of two types (a) intrinsic resistance (b) the resistance due to imperfection. In a perfect crystal the intrinsic resistance is due to the periodic variation of strain energy as dislocations move through the crystal. But in real crystals most of the resistance arises from imperfections that act as obstacles to dislocation motion. The relation between hardness and inter atomic distance is given by (WOLF et al.)

$$H_v = \text{Constant } Xr^{-m}\,, \tag{2}$$

where r is the inter atomic distance and m is a constant, its value depending upon the structure of a particular system. This correlation of hardness with inter atomic distance is confined to pure crystals only. But in order to correlate the hardness of mixed crystals to the inter atomic distance the Lattice parameters of KCl–RbCl mixed crystals as shown in Table 1 are used to calculate the microhardness and shown in curve b of Figure 2 which appears to be a straight line. But experimentally it is observed that the microhardness varies non-linearly with composition as represented by curve (a) Figure 2. To accout for this discrepancy a disorder parameter can be included in the relation so that

$$H = \text{Const } Xr^{-m} + KN_aN_b\,, \tag{3}$$

where N_a and N_b are molar concentrations of the components and K being the constant. The values of m and K are found to be 6 and 41 kg/mm² for these crystals and the disorder parameter is comparatively lower in this system than the other two systems of KCl–KBr and KBr–KI (SUBBA RAO, HARI BABU). The calculated values are in good agreement with experimentally observed values as shown by curves (a) and (c) in Figure 2. Thus the expressions for microhardness has two terms one connected to intrinsic nature depending upon structure insensitive parameter and the

Table 1

% molar concentration of RbCl in KCl	lattice parameter a/Å	ionic radii Å
0	6.2857	3.142
6	6.2957	3.15
12	6.3267	3.163
18	6.3458	3.172
25	6.3708	3.185
40	6.4012	3.20
50	6.4434	3.22
63	6.4796	3.24
87	6.5465	3.273
100	6.5919	3.295

Table 2

Mixed System	% of difference in lattice parameter	difference in ionic radii	H_v max
KCl–RbCl	4.5[c]	0.15	8.77
KCl–KBr	4.8[c]	0.14	14.7[b]
NaCl–NaBr	5.5[c]	0.14	22.86[b]
KBr–KI	7.1[c]	0.21	24.1[a]

[a]) BHIMA SHANKARAM: Ph. D. thesis, Osmania University, Hyderabad, India
[b]) SUBBA RAO, HARI BABU
[c]) TOBOLOSKY, A. V.: J. Chemical Physics 10, 187 (1942)

other connected with the disorder parameter depends upon the concentration of imperfections, the impurities, impurity vacancy pairs, dislocations etc. The dislocations are said to be created, when the lattice experiences strain due to misfit of substitutional ion. The ionic radius of K^+ ion is 1.33 Å where as that of Rb^+ ion is 1.48 Å. The percentage difference of lattice parameter is 4.5. Hence KCl and RbCl form a solid solution of complete miscibility than any other solid solution of alkali halide mixed crystals at room temperature. Rb^+ ions are randomley distributed in the solid solution occupying the substitutional positions. So when a Rb^+ ion of larger size enters the lattice, then lattice will experience strain to accommodate. Due to these internal strains more dislocations are created. From Table 2 it can be observed that the percentage of difference in lattice parameter of KCl–RbCl system is smaller than the other mixed systems. Similarly the percentage difference in ionic radii is also smaller. Hence, the increased maximum variation in hardness is smaller in the present system than the other two systems.

Thus it can be concluded that when a solid solution of complete miscibility is formed due to substitutional replacement of ion of different size, lattice strains develop which contribute to more imperfections leading to increase in microhardness of mixed crystals than pure ones.

The authors are thankful to the Head, Dept. of Physics O.U. for his constant encouragement and authors ABR and KJP are thankful to UGC. New Delhi for providing Teacher Fellowship.

References

JOHNSTON, W. G.: J. Appl. Physics 33, 2050 (1962)
LAKSHMIPATHI RAO, M. and HARI BABU, V.: Ind. J. Pure Appl. Physics 16, 821 (1978)
MEYER, E.: V.D.I. 52, 645 (1908)
PRATAP, K. J., HARI BABU, V.: Bull. Mater. Sci. 43 (1980)
SUBBA RAO, U. V., HARI BABU, V.: Pramana Vol. 11, 149 (1978)
UPIT, G. P. & VARCHENYA, S. A.: phys. stat. sol. 17, 831 (1966)
WOLF, G. A., TOMAN, L., FIELD, N. I., CLARK, J. C.: Halbleiter and phosphore, New York 1958 p. 463
YOST, F. G.: Metallurgical transactions 14A, 947 (1983)

(Received February 24, 1987)

Authors' address:
Prof. V. HARI BABU, K. J. PRATAP, A. BALARAMAIAH
Department of Physics, University College of Science, Osmania University
Hyderabad — 500007
India

Book Review

R. J. D. TILLEY

Defect Crystal Chemistry and its Applications

Blackie and Sons Limited, Glasgow and London. Published in USA by Chapman and Hall, New York, 1987, 236 Seiten, 16 Tabellen, 94 Abbildungen, Preis: £ 35.00, ISBN 0-412-01331-2

Das Buch ist aus Vorlesungen und Übungen für Chemiker, Physiker und Werkstoffwissenschaftler in fortgeschrittenen Abschnitten des Studiums und Weiterbildungskursen hervorgegangen. Es knüpft in 10 Kapiteln, die zu 4 Abschnitten zusammengefaßt sind, an moderne Fragestellungen aus der gegenwärtig sehr dynamisch verlaufenden Entwicklung der Festkörperforschung an, ohne auf grundlegende Zusammenhänge zu verzichten. Diese werden unter Beschränkung auf den wesentlichen Gehalt prägnant entwickelt und als methodisches Instrument unmittelbar auf strukturelle Fragen oder eine Kombination von Eigenschaften und deren praktische Nutzung angewendet.

Im ersten Abschnitt stehen die klassischen Fehlstellenbildungsgleichgewichte und das Bewegungsverhalten atomarer Teilchen im Festkörper im Mittelpunkt. Die Einführung der Punktdefekte wird im ersten Kapitel mit dem photographischen Prozeß und der LiI-Batterie verbunden, die phänomenologische und die atomistische Beschreibung der Transportvorgänge im zweiten und dritten Kapitel eingehend vorgenommen, wobei auch hier jeweils Beispiele zur Erläuterung dienen. Der zweite Abschnitt ist mit den Kapiteln 4 und 5 den nichtstöchiometrischen Materialien mit definierter Valenz, deren Strukturen und Anwendungen vorbehalten. Hier wird die Fehlstellensymbolik eingeführt und die Chemie der Defekte erläutert. Durch Eliminierung eines Bestandteils, z.B. von Sauerstoff aus Oxiden, verursachte Nichtstöchiometrie und ihre Indikation durch Elektronenleitung werden im dritten Abschnitt Kapitel 6 behandelt. Springmechanismen, Bandleitung und thermoelektrische Effekte sind ebenso einbezogen wie die Farbzentrenbildung und ihre Bedeutung für die Informationsaufzeichnung im Kapitel 7. Der sorgfältigen Erklärung einer richtigen thermodynamischen Behandlung der Phänomene ist Kapitel 8 gewidmet. Das Buch endet im 4. Abschnitt mit der Entwicklung typischer Strukturen nichtstöchiometrischer Phasen, Cluster und Blockstrukturen mit eindrucksvollen Beispielen der mittels hochauflösender Elektronenmikroskopie erreichten Aussagen (Kapitel 9). Am Schluß, das 10. Kapitel geht dem interessanten Zusammenhang Thermodynamik und Struktur nach.

Den Kapiteln sind vielfach zusammenfassende Wertungen mit Schlußfolgerungen nachgestellt sowie weiterführende Literatur. Die Monographie gibt eine Einführung, erklärt exemplarisch, das regt zum selbständigen Übertragen auf eigene Probleme an. Sie ist von hohem didaktischem Wert und wird sich in Fachbibliotheken wie im unmittelbaren persönlichen Gebrauch rasch einführen.

A. FELTZ

Cryst. Res. Technol.	22	1987	9	1211–1214

K. R. K. RAO, J. V. RAO, P. VENKATACHARYULU,
S. V. CHIRANJEEVI RAO, V. BALIAH

Physics Department, Nagarjuna University, Nagarjunanagar, India

Phase Transition Studies of the Liquid Crystal 4-[4-n-Hexyloxybenzylidene amino] Azobenzene Thermal Microscopy and DSC

The phase transition temperatures of the liquid crystal 4-[4-n-hexyloxybenzylideneamino]-azobenzene have been investigated by the thermal microscopy and the Differential Scanning Calorimetric (DSC) studies. These studies confirm that this mesogen exhibits nematic, smectic A and smectic B phases. The transition enthalpies, transition entropies and also the textures exhibited by the different phases are presented.

Introduction

The various phases and the phase transition temperatures that in general exhibited by the various mesogens can be well identified and confirmed by the well established techniques like the Thermal Microscopy and the Differential Scanning Calorimetric studies. However, the DSC studies enables one also to determine the transition enthalpies and entropies. It has been well explained by SACKMANN and DEMUS that the phase classification of a whole homologous series is possible if one or several members of the series are classified and the transition temperatures are known. Previously the phase classification and the transition temperatures pertaining to the whole homologous series of the 4-[4-n-alkoxybenzylidene amino] azobenzenes starting from methyl and up to octadecyl was given by DEMUS et al. The classification of the phases relating to the nonyl and dodecyl (DEMUS et al.) members of the above series were classified earlier, from textural and miscibility studies and the transition temperatures relating to these two members were given by Arora and Fergason and also by FISHEL and PATEL. Further DIELE et al. confirmed the structure of the liquid crystalline phases of the nonyl member of these series by X-ray investigations. As a part of our systematic studies on the phases and phase transition temperatures of mesogens (RAO et al.), in one of our earlier papers, we reported the phases and phase transition temperatures relating to the liquid crystal 4-[4-n-Pentyloxybenzylidene amino] azobenzene from thermal microscopy and DSC studies and we now present here the results of our investigations that we have carried out on the 4-[4-n-Hexyloxybenzylidene amino]-azobenzene. Eventhough our present observations in this compound relating to the phase transition temperatures are in agreement with those reported by FISHEL and PATEL, the characterisation of the phases in the present studies are in disagreement with those reported by FISHEL and PATEL and the present characterisation is also supported by our DSC studies.

Experimental

4-[4-n-hexyloxybenzylidene-amino]-azobenzene (H_xBAAB) was prepared by condensation of equimolar quantities of p-aminoazobenzene and p-hexyloxybenzaldehyde in refluxing ethanol in the presence of a few drops of acetic acid. The solvent was removed by distillation. The compound was recrystallised from ethylacetate until the phase transition temperatures are found to be constant.

The textural studies were carried out by a polarising microscope with hot stage arrangement. The phases exhibited by H_xBAAB are presented below:

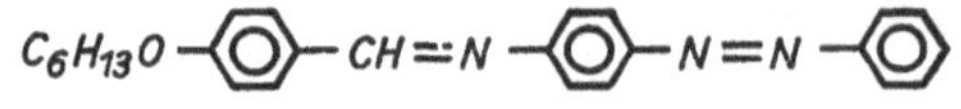

4-[4-n-hexyloxybenzylidene amino] azobenzene

Solid ↔ S_B ↔ S_A ↔ N ↔ I

DSC scans were obtained from Perkin Elmer 1B differential scanning calorimetry. The DSC scans were run at a rate of 8 K per minute.

Results and discussion

On cooling the isotropic liquid, the marble texture is observed at 445 K which is the indication of the onset of nematic phase. The marble texture continues till the temperature reaches to 420 K where the marble texture changes to focal conic fan texture (DEMUS, RICHTER). This indicates that the observed phase is smectic A. Our finding is in agreement with that of ARORA and FERGASON and also that of FISHEL and PATEL. On further cooling the temperature to 398 K transition bars (SMITH, GARDLUND) were observed across the fans. The transition bars were observed for a temperature interval of 0.1 or 0.2 K and once again the focal conic fan texture with discontinuities is observed. The above facts indicate that the observed phase is smectic B. After keeping the temperature constant for several hours the focal conic fan texture of smectic B phase changes to mosaic texture (DEMUS, RICHTER) of Smectic B phase which is the natural texture of Smectic B phase. ARORA and FERGASON reported that H_xBAAB exhibits smectic B phase but Fishel and Patel reported that this phase in H_xBAAB

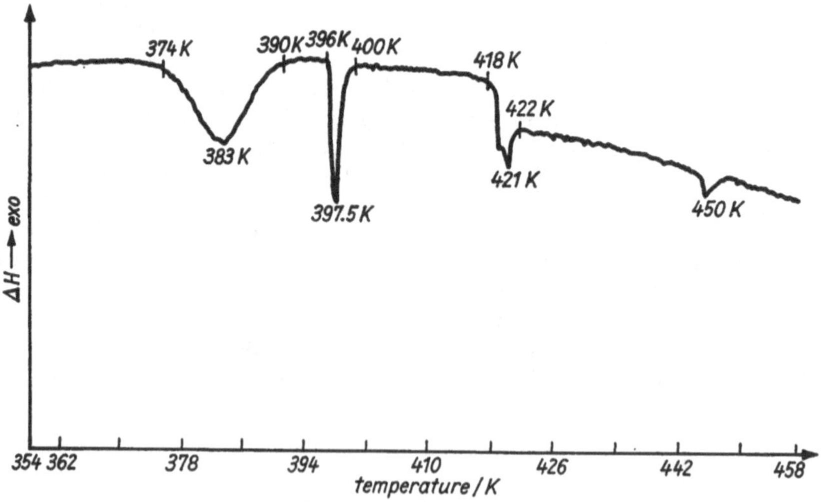

Fig. 1. DSC scan for H_xBAAB

is smectic C but from the present studies it is confirmed that this phase is smectic B and we are in agreement with ARORA and FERGASON.

The DSC scans are presented in Figure 1. The phase transition temperatures, transition enthalpies and transition entropies at various phase transitions are presented in Table 1 along with the thermal microscopic values. A very low value in entropy is observed across N–I transition which may be caused by the limit of error of the instrument as the instrument causes remarkable experimental errors especially in the case of low enthalpy values. However, there are slight deviations in the phase transition temperatures as observed from the thermal microscopy and DSC studies.

Table 1
Transition Temperatures, enthalpies and entropies for H_xBAAB

Transition	transition temperature K	ΔH J/mole	ΔS J/mole · K^{-1}
$C-S_B$	383*) 377**)	6795	18
S_B-S_A	397.5*) 398**)	2254	6
S_A-N	421*) 420**)	1307	3
$N-I$	450*) 445**)	16	0.04

*) Transition temperatures from DSC studies
**) Transition temperatures from thermal microscopy

In DSC studies also a definite change in enthalpy at S_A–S_B transition is observed. If the transition is from S_A to S_C there should not be any change in enthalpy because of the minimum structural change involved at S_A to S_C transition. Hence from the DSC studies also it can be confirmed that the transition in H_xBAAB at 398 K is from smectic A to smectic B but not from smectic A to smectic C. Hence H_xBAAB exhibits nematic, smectic A and smectic B phases.

K. R. K. Rao and J. V. Rao thank the University Grants Commission, New Delhi for financial assistance.

References

ARORA, S. L., FERGASON, J. L.: Symp. Faraday Soc. No. 5, 97 (1971)
DEMUS, D., SACKMANN, H.: Z. Phys. Chem. (Leipzig) **238**, 215 (1968)
DEMUS, D., KUNICKE, G., PELZL, G., ROHLING, B., SACKMANN, H., SLAFFNER, R.: Z. Phys. Chem. (Leipzig) **254**, 373 (1973)
DEMUS, D., DEMUS, H., ZASCHKE, H.: Flüssige Kristalle in Tabellen, 2nd Edition, Leipzig 1974
DEMUS, D., RICHTER, L.: Textures of Liquid Crystals, Weinheim, New York 1978
DIELE, S., BRAND, P., SACKMANN, H.: Mol. Cryst. Liq. Cryst. **16**, 105 (1972)
FISHEL, D. L., PATEL, P. R.: Mol. Cryst. Liq. Cryst. **17**, 139 (1972)
RAO, J. V., MURTY, C. R. K.: Phase Transitions **5**, 67 (1985)
RAO, J. V., CHOUDARY, L. V., MURTY, C. R. K., RAO, K. R. K.: Phase Transitions **5**, 72 (1985)

Rao, K. R. K., Rao, J. V., Choudary, L. V., Murty, C. R. K.: Acustica **57**, 98 (1985)
Rao, K. R. K., Rao, J. V., Venkatacharyulu, P., Baliah, V.: phys. stat. sol. (a) **93**, 93 (1986)
Sackmann, H., Demus, D.: Mol. Cryst. **2**, 81 (1966)
Smith, G. W., Gardlund, Z. G.: J. Chem. Phys. **59**, 3214 (1973)

(Received 1) September 20, 1986;
2) February 13, 1987)

Authors' addresses:

Dr. P. Venkatacharyulu, J. V. Rao
Physics Department, Nagarjuna University
Nagarjunanagar 522510
India

S. V. Chiranjeevi Rao
IDL Chemicals, Hyderabad, Andhra Pradesh, India

V. Baliah
Fromer Vice-Chancellor, Nagarjuna University
K. R. K. Rao
Chemistry Department, Nagarjuna University

Cryst. Res. Technol.	22	1987	9	1215–1221

P. VENKATACHARYULU, A. V. N. GUPTA, V. SUBBA RAO, J. V. RAO

Department of Physics, Nagarjuna University, Nagarjunanagar, India

Phase Transition Studies in Certain Thermotropic Liquid Crystals

Thermal Microscopy and DSC

Investigations on the phase transition studies in the thermotropic liquid crystals, (1) p-cyanophenyl trans-4-propylcyclohexane carboxylate, (2) p-cyanophenyl trans-4-pentylcyclohexane carboxylate, (3) 4′-n-pentyloxy-4-cyanobiphenyl and (4) 4′-n-octyloxy-4-cyanobiphenyl are carried out by the thermal microscopy and the differential scanning calorimetric (DSC) studies. These studies reveal that the first three mesogens exhibit a single mesophase while the fourth one exhibits a smectic A phase in addition to the nematic phase. The textures exhibited by these mesogens in their mesophases and their transition temperatures, enthalpies and entropies are presented.

Introduction

Investigations on liquid crystalline phase transitions are the subject of general interest now-a-days for chemical physicists as they exhibit different mesophases in between two extremes namely the solid crystalline and the isotropic liquid states. The various phases and their phase transition temperatures that in general exhibited by the various mesogens can be well identified and confirmed by the well established techniques like the thermal microscopy and the differential scanning calorimetric (DSC) studies. The thermal microscopy enables one to investigate the various textures exhibited by different liquid crystals in their liquid crystalline phases in addition to the determination of their phase transition temperatures, whereas the DSC studies enables one to determine the transition enthalpies and entropies besides the determination of the transition temperatures as well. It has been well explained by SACKMANN and DEMUS that the phase classification of a whole homologous series is possible if one or several members of the series are classified and the transition temperatures are known.

The transition temperatures and the various phases that are present in two liquid crystal esters namely (1) p-cyanophenyl trans-4-propylcyclohexane carboxylate and (2) p-cyanophenyl trans-4-pentylcyclohexane carboxylate and also two 4′-n-alkoxy-4-cyanobiphenyls namely (1) 4′-n-pentyloxy-4-cyanobiphenyl and (2) 4′-n-octyloxy-4-cyanobiphenyl (80CB) have been reported by different earlier workers (M/s Hoffmann-La-Roche; COX, JOHNSON; LESNIAK et al.; SUSHMITA SEN et al.). No detailed investigation on the textures exhibited by different phases by these mesogens in their liquid crystalline state is reported so far. There is some controversy regarding the presence of Smectic A phase in 80CB as some earlier investigators (M/s Hoffmann-La-Roche; SUSHMITA SEN et al.) did not identify the same while some others (COX, JOHNSON, LESNAIK et al.) reported the existence of the same. So far no DSC studies were reported on the two liquid crystal esters, namely, p-cyanophenyl trans-4-propylcyclohexane carboxylate and p-cyanophenyl trans-4-pentyl cyclohexane carboxylate.

As such in the present studies we report the results of the thermal microscopic and differential scanning calorimetric studies that we have carried out on the above mentioned four liquid crystals and our studies also confirm the presence of smectic-A phase in 8OCB.

Experimental

All the four above mentioned liquid crystals used in the present studies were imported from M/s Hoffmann-La-Roche, Switzerland, under the trade names liquid crystal components 1830, 1850, 5315, and 5324, in their pure form. The textural studies and their transition temperatures were carried out by a polarizing microscope with a hot stage arrangement. The various phases exhibited by the different liquid crystals are as follows:

1. *p-cyanophenyl trans-4-propylcyclohexane carboxylate (1830)*

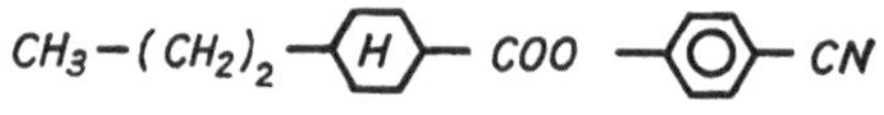

Solid ⟷ Nematic ⟷ Isotropic

2. *p-cyanophenyl trans-4-pentylcyclohexane carboxylate (1850)*

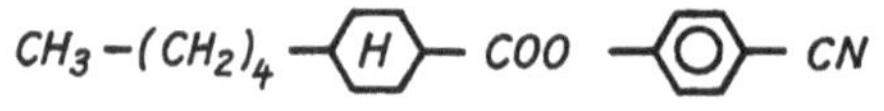

Solid ⟷ Nematic ⟷ Isotropic

3. *4'-n-pentyloxy-4-cyanobiphenyl (5OCB) (5315)*

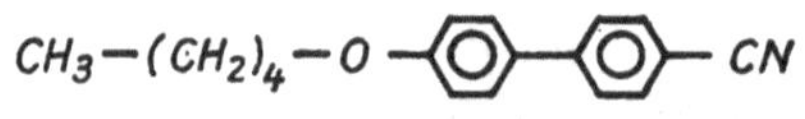

Solid ⟷ Nematic ⟷ Isotropic

4. *4'-n-octyloxy-4-cyanobiphenyl (8OCB) (5324)*

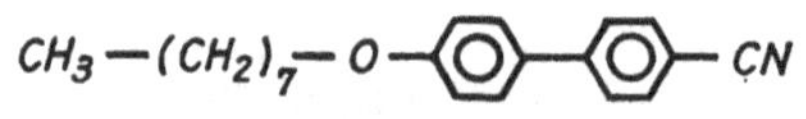

Solid ⟷ Smectic A ⟷ Nematic ⟷ Isotropic

The DSC scans were obtained from Perkin Elmer 1B differential scanning calorimetry. The scans were run at a rate of 8 K per minute.

Results and discussion

A. Thermal microscopy:

The phase transition temperatures and the range of the liquid crystalline phases for all the above mentioned liquid crystals are indicated in Table 1, along with the literature values (M/s Hoffmann-La-Roche; COX, JOHNSON, SUSHMITA SEN et al.; LESNAIK et al.). In the first three liquid crystals only a single liquid crystalline phase, i.e.,

Table 1
Transition Temperatures of the Various Mesogens

Investigators	mesogen	C – S_m/N K	S_m – N K	N – I K
present Studies	1830	326.5 (N)	–	342.3
(thermal microscopy)	1850	320.0 (N)	–	351.7
	5315 (5OCB)	320.5 (N)	–	340.5
	5324 (8OCB)	327.5 (S_mA)	338.8	352.2
Hoffmann-La-Roche	1830	326.5 (N)	–	342.5
	1850	320.0 (N)	–	351.7
	5315 (5OCB)	320.5 (N)	–	340.5
	5325 (8OCB)	327.5 (N)	–	352.3
Cox and JOHNSON	5315 (5OCB)	319.02 (N)	–	339.61
	5324 (8OCB)	326.0 (S_mA)	338.58	352.03
LESNAIK et al.	–	–	–	–
	5324 (8OCB)	235.9 (S_mA)	338.7	355.6
SUSHMITA SEN et al.	5315 (5OCB)	231.0 (N)	–	340.5
	5324 (8OCB)	340.0 (N)	–	353.0

nematic phase is observed in between the solid crystalline and isotropic liquid states. All the three liquid crystals melt from solid crystal to nematic phase at the temperatures given in Table 1 and exhibit marble texture which is characteristic of the nematic phase (DEMUS, RITCHER). These liquid crystals exhibit these marble textures until they attain their respective nematic-isotropic transition temperatures. The textures are observed both in the heating and cooling cycles at the same transition temperatures.

The fourth liquid crystal 8OCB exhibits a smectic A phase in addition to the nematic phase in between the solid crystalline and isotropic states, with its transition temperatures also indicated in the same Table 1. This compound melts from the solid state to the Smectic A phase at 327.5 K. In this phase it exhibits focal conic fan texture which is characteristic texture of Smectic A phase (DEMUS, RITCHER). This texture continues until the temperature is rised to 338.8 K, when the texture transforms from focal conic to marble texture which is an indication of the change of phase from smectic A to nematic (DEMUS, RITCHER) and this texture continues until 352.2 K when the compound becomes isotropic liquid. In the cooling cycle, on cooling the isotropic liquid the marble texture which is characteristic of the nematic phase is seen again, when the mesogen attains 352.2 K. On further cooling to 338.8 K the batonnets (DEMUS, RITCHER) are formed which later coalesce to give a focal conic fan texture which is an indication of the Smectic A phase. This smectic A phase continues till the sample is cooled to 327.5 K.

In this liquid crystal a single mesophase, nematic phase was reported by SUSHMITA SEN et al. and also by the manufactures of this mesogen namely M/s. Hoffmann-La-Roche, Switzerland. But Cox and JOHNSON in their DSC studies and also LESNAIK et al. in their conductivity measurements also observed smectic A phase in addition to the nematic phase. Our present studies also confirm the presence of smectic A phase in this liquid crystal in addition to nematic phase.

B. DSC Studies

The DSC scans for these four mesogens are presented in Figures 1, 2, 3, and 4. The phase transition temperatures, transition enthalpies and transition entropies in all

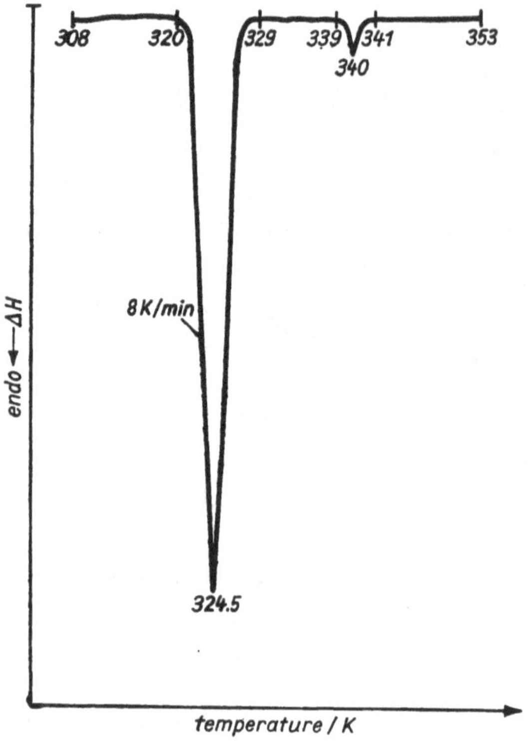

Fig. 1. DSC scan for p-cyanophenyl trans-4-propylcyclohexane carboxylate

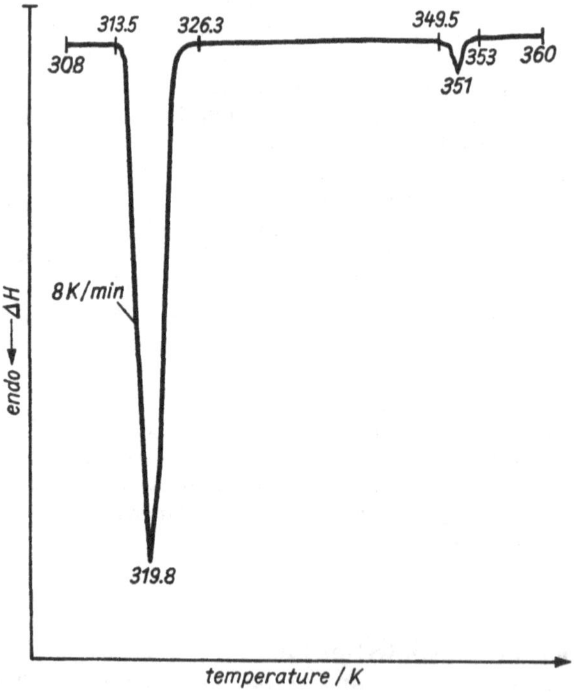

Fig. 2. DSC scan for p-cyanophenyl trans-4-pentylcyclohexane carboxylate

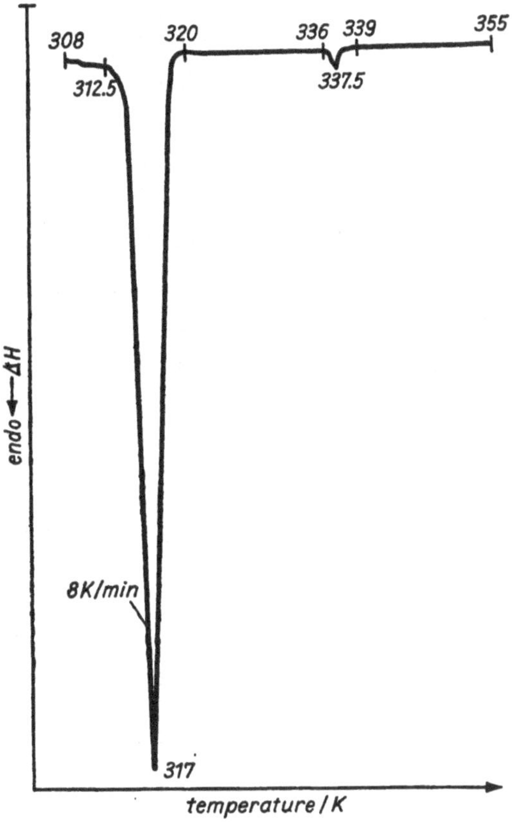

Fig. 3. DSC scan for 4'-n-pentyloxy-4-cyanobiphenyl

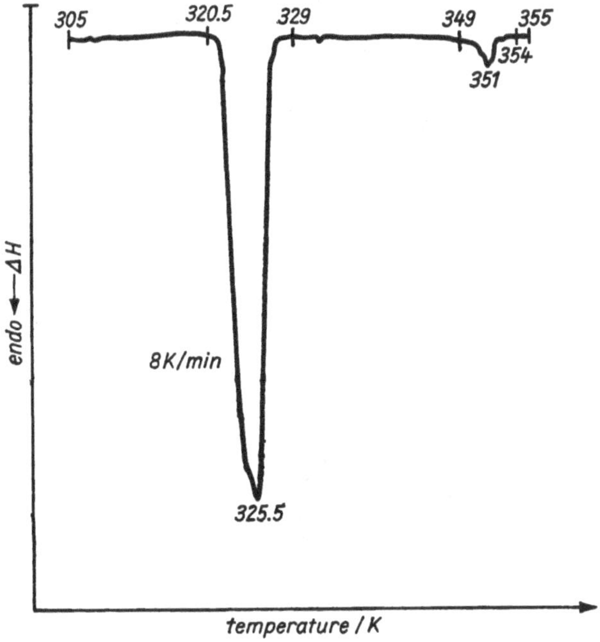

Fig. 4. DSC scan for 4'-n-octyloxy-4-cyanobiphenyl

Table 2

Thermodynamic Data on 1830, 1850, 5315, 5324

Liquid crystal component	transition	transition temperature K			ΔH cal/mole			ΔS cal/mole · K^{-1}	
		this work	Cox and JOHNSON	GRAY et al.	this work	Cox and JOHNSON	GRAY et al.	this work	Cox and JOHNSON
1830	C—N	324.5	—	—	5450	—	—	16.8	—
	N—I	340.0	—	—	122.16	—	—	0.36	—
1850	C—N	319.8	—	—	6196	—	—	19.37	—
	N—I	351.0	—	—	164	—	—	0.467	—
5315	C—N	317.0	319.02	321.0	6248	62.39	6900	19.7	19.54
				326.0					
	N—I	337.5	339.61	340.5	89	71.7	—	0.26	0.211
5324	C—S_A	325.5	326.0	327.5	6196	7288	5900	19.37	22.3
	S_A—N	—	338.58	340.0	—	18.7	—	—	0.055
	N—I	351.0	352.03	353.0	164	235	—	0.467	0.668

these four compounds are presented in Table 2 along with the values available in literature for the liquid crystals 5OCB and 8OCB for comparison. However, there are slight deviations in the phase transition temperatures as observed from the thermal microscopy and DSC studies.

In 4′-n-octyloxy-4-cyanobiphenyl the DSC scan did not indicate the presence of smectic A-nematic phase transition, whereas our thermal microscopic studies indicate the presence of the same. Cox and Johnson observed this transition in their DSC studies. Lesnaik et al. also observed the same in their anisotropic electrical conductivity measurements. The reason for not observing this transition in the present DSC studies may be due to the fact that the energy involved in this transition is very less and it is a weak first order transition tending to the second order, as our specific volume studies indicate (Venkatacharyulu et al.), which might have been less than the exiting energy of the instrument.

The first two liquid crystals, i.e., p-cyanophenyl trans-4-propylcyclohexane carboxylate and p-cyanophenyl-trans-4-pentylcyclohexane carboxylate belong to the homologous series of p-cyanophenyl-trans-4-alkylcyclohexane carboxylates and the two liquid crystals 5OCB and 8OCB belong to the homologous series of p′-n-alkoxy-p-cyanobiphenyls, show an increasing trend in the transition temperature with increasing molecular weights or chain lengths which is a general trend. All the transitions, i.e., crystal-mesophase and mesophase-isotropic transitions are found to be of first order in these mesogens except smectic A-nematic transition observed in 8OCB can be a weak first order tending to second order. This conclusion is confirmed from the measurements of Cox and Johnson (Tab. 2) as they obtained very low values of ΔH and ΔS at this transition when compared to the values obtained at other transitions in 8OCB.

The authors are deeply indebted to Prof. A. Szymanski, Institute of Mathematics and Physics, I. Łukasiewicz Technical University, Rzeszov, Poland for his kind gesture in supplying 8OCB as a gift sample which enabled to confirm the results obtained in the sample procured from M/s Hoffmann-La-Roche, Switzerland. The authors are also thankful to IDL Chemicals, Hyderabad, India for recording the DSC scans. One of the authors (AVNG) is thankful to the authorities of T.J.P.S. College for their encouragement.

References

Cox, R. A., Johnson, J. F.: IBM J. Research and Development **22** (1978)

Demus, D., Richter, L.: Textures of Liquid Crystals, Weinheim, New York 1978

Gray, G. W., Harrison, K. J., Nash, J. A., Constant, J., Hulme, D. S., Korton, J., Raynes, E. P.: Proceedings of the Symposium on Ordered Fluids and Liquid Crystals, R. S. Porter and J. F. Johnson, eds, New York 1974

Lesnaik, M., Szwajczak, E., Labno, W., Szymanski, A.: Advances in Liquid Crystal Research and Applications, Oxford 1980

Sackmann, H., Demus, D.: Mol. Cryst. **2**, 81 (1966)

Sushmita Sen (Miss), Pradip Brahma, Subir, K. Roy., Mukherjee. D. K., Roy, S. B.: Mol. Cryst. Liq. Cryst. **100**, 327 (1983)

Venkatacharyulu, P., Gupta, A. V. N., Rao, J. V., Prasad, K. S. R., Prasad, N. V. L. N.: Phase Transitions (communicated)

(Received February 13, 1987)

Authors' address:

Dr. P. Venkatacharyulu, A. V. N. Gupta, V. Subba Rao, J. V. Rao
Department of Physics, Nagarjuna University
Nagarjunanagar 522510, India

Book Review

A. Krawietz

Materialtheorie — mathematische Beschreibung des phänomenologischen thermomechanischen Verhaltens

Berlin – Heidelberg – New York – Tokyo, Springer-Verlag 1986, 476 S., 83 Abb., 3 Tab, DM 140,–, ISBN 3-540-15435-3

Die phänomenologische Materialtheorie ist heute ein wichtiges Hilfsmittel für Entwurf, Fertigung und Betrieb von technischen Konstruktionen, aber auch zur Beschreibung natürlicher Stoffe, wie sie z. B. in der Geowissenschaft oder der Medizin vorkommen. Die Berücksichtigung von Temperaturänderungen und großen Verformungen erfordert dabei eine einheitliche Behandlung der kinematischen, dynamischen und thermodynamischen Grundlagen des Materialverhaltens sowie der allgemeinen Regeln zur Aufstellung von Materialgleichungen. Dieser Problematik ist ein wesentlicher Teil des vorliegenden Buches gewidmet. Zum leichteren Verständnis der anspruchsvollen Theorie werden einleitend rheologische Modelle vorangestellt, die auch der Festlegung eines auf diskreten inneren Variablen beruhenden Zustandsbegriffes und der späteren ausführlichen Diskussion von konkreten Materialeigenschaften wie Elastizität, Viskosität, Plastizität und deren Kombinationen dienlich sind. Unter der Fülle des angebotenen Stoffes finden sich Hinweise auf solche wichtigen, in Kontinuumsmechaniklehrbüchern bisher wenig behandelten Probleme wie effektive Theorie heterogener Materialien, Phasenübergänge, Memory-Effekt und die, allerdings etwas kritischer zu bewertende kinematische Aufspaltung der Deformation in elastische und inelastische Anteile. Das Buch ist durch eigene Arbeiten des Autors zum Konzept der Materialpassivität geprägt, welches in speziellen Situationen eine entropiefreie Formulierung des zweiten Hauptsatzes gestattet, ein nach Meinung des Rezensenten sehr plausibler Standpunkt. Jedoch wird der Leser möglicherweise etwas verwundert sein, die Thermodynamik auf eineinhalb Seiten abgehandelt zu sehen und dort nichts über innere Energie und den ersten Hauptsatz zu finden. Das Buch enthält auch Ausführungen zur Materialstabilität. An dieser Stelle wäre eine schärfere Unterscheidung zwischen Stabilität von Materialien und Körpern einschließlich Umgebung wünschenswert.

Die verwendete mathematisch abstrakte Darstellung der konstitutiven Theorie geht unter Hinzunahme eines komprimierten Abschnittes über lineare Algebra auf weniger bekannte Arbeiten Nolls zurück. Das dadurch erwartete bessere Verständnis des physikalischen Inhaltes erfordert jedoch sicher noch etwas Übung bei den Lesern. Insofern wird das reichhaltige Buch neben theoretisch interessierten Studenten und Ingenieuren hauptsächlich Spezialisten der Kontinuumsmechanik zu empfehlen sein. H. Balke

Cryst. Res. Technol.	22	1987	9	K141–K144

Short Notes

K. SOMAIAH

Department of Physics, Osmania University, Hyderabad, India

Dislocation Activity in BaFCl Single Crystals

Introduction

The method which gives a quantitative estimation of defect content in the materials, is the study of dislocations by etching process. The process of selective etching has been widely used in recent years for revealing the sites of emergence of dislocations of the surface of crystals. Upto date an extensive information is available about selective etching and dissolution of ionic crystals. But no systematic investigation of these processes in BaFCl crystals has been reported. However, relatively little work has been done to characterize the nature of dislocations in BaFCl crystals (PATEL, SINGH; BHAT, NARAYANAN; SOMAIAH, HARI BABU 1979, 1983). The results of a systematic investigation of selective etching and dissolution of BaFCl crystals are reported in this paper. The action of the etchant on (001) plane under different conditions has been analysed.

Experimental

The single crystals used in the present investigation were grown by flux technique using $BaCl_2$ as common material and BaF_2 or NaCl being as flux. Thus two types of crystals are obtained grown from two different fluxes. Before etching, the crystals were cleaved parallel to the (001) or perpendicular to the crystallographic C-axis. In our initial work, chemical etching was performed using water and formic acid independently at room temperature (SOMAIAH, HARI BABU 1983). It was found that water etchant acted very quickly whereas the formic acid etchant acted very very slow. This prompted a search for an etchant yielding pits in a reasonable time. Thus an etchant consisting of 50% water and 50% formic acid is found to give the pits in a reasonable time (hereafter called FW etchant). The etch pits were observed using an Nu2 microscope (Carl Zeiss, Jena).

Results and discussion

Though all the three etchants i.e., formic acid, water and FW etchant could reveal the etch pits on the (001) faces of the BaFCl crystals, water etchant provided a greater resolution to reveal microstructural features.

In the present study (Fig. 1a), the etch pits on (001) were found to be square with their edges parallel to the $\langle 100 \rangle$ directions. The size of the pits and their density varied from about $10 \sim 30\ \mu m$, and $10^3 - 10^5 \cdot cm^{-2}$, respectively. The pits occurred either individually (in random distribution) or in linear arrays along the $\langle 100 \rangle$ direction. The linear arrays of etch pits are interpreted as resulting from dislocations on

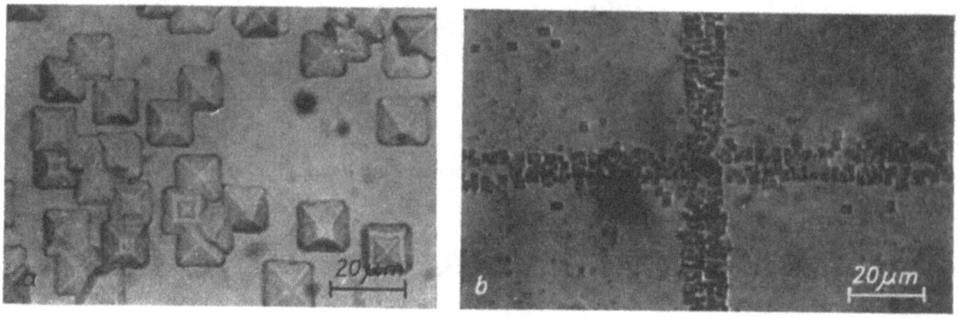

Fig. 1. Dislocation etch pits obtained on the surface of BaFCl crystal using FW etchant, etching time 5 minutes

slip planes that intersect the (001) surface in ⟨100⟩ direction. Figure 1b shows a commonly occurring feature. Pairs of linear arrays are seen extending from centrally located defects which appear as darker spots at the locations of the intersections of the arrays. These darker spots generally coincide with intersections of the curve-linear distributions of a micrometer-sized inclusions that pass through the interior of the crystal. Such inclusions have been already reported in other crystals (ROZHANSKI et al.).

In order to investigate the specific nature of the linear arrays of etch pits visible in Figure 1b, the effect of deformation by microhardness indentation was examined. In particular it was sought to determine whether or not such arrays could consist of mobile dislocations resulting from crystallographic slip. Figure 2 shows the results of these investigations. Figure 2a shows the (001) plane of a freshly cleaved crystal after deforming with a 10 g Vickers microhardness indentation. Figure 2b shows linear arrays of etch pits protruding in ⟨100⟩ direction from the indentation. These etch pits can be interpreted as dislocations that were introduced by the indentation, i.e. plastic deformation and are present on the slip band that intersect the (001) planes in the ⟨100⟩ type directions. Similar conclusions were drawn from etching and deformation studies on single crystals of MgO, LiF, and NH_4Cl, whereas these results do not show conclusively that the linear arrays of defects are mobile dislocations (KOSTIN et al.). Our observations have also shown that aged and fresh dislocations (formed by local deformation) etch differently. On the aged dislocations the etch pits are much deeper.

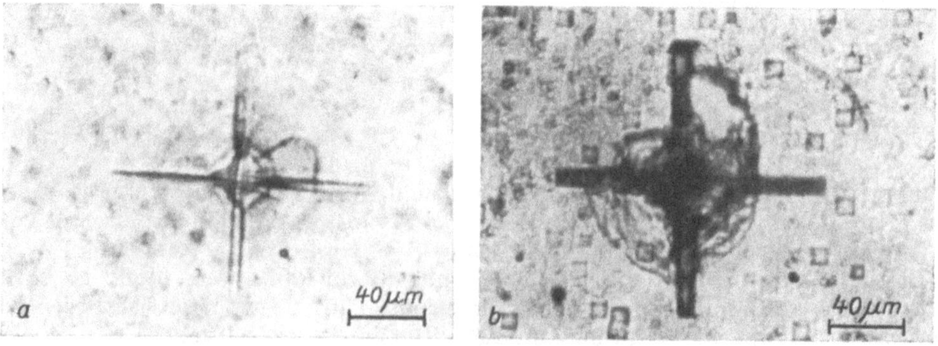

Fig. 2. Vickers indentation on the freshly cleaved surface of BaFCl crystal before etching (a) and after etching for 2 minutes in FW etchant (b)

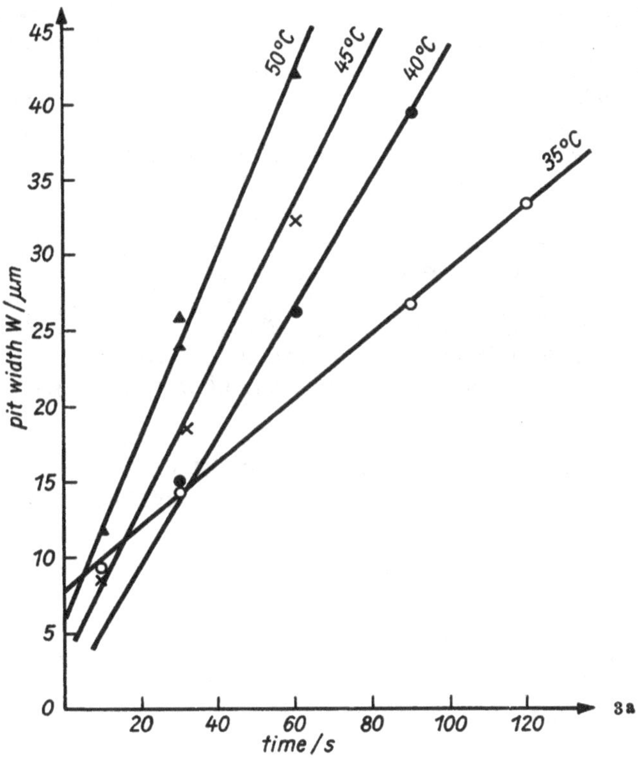

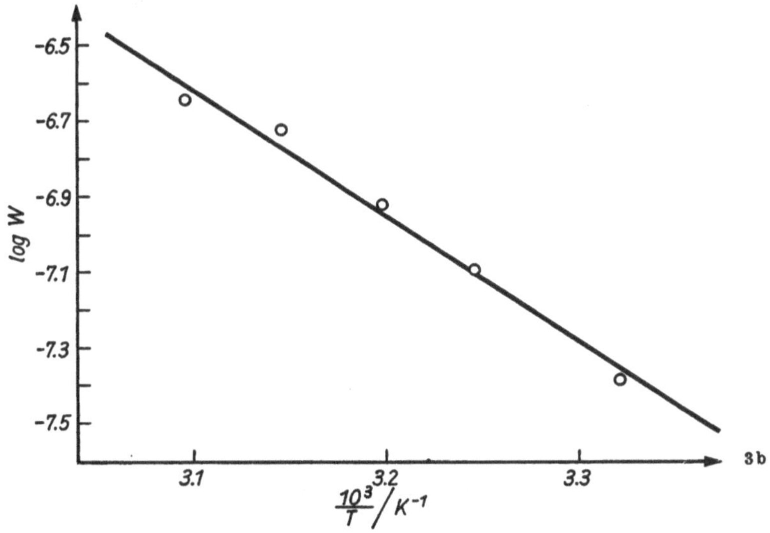

Fig. 3. Variation of pit width with time of etching at different temperatures (a) log of etch pit widening rate W against the reciprocal of temperature (b)

In order to calculate the activation energy E_t for the tangential movement of step away from the dislocation, etch pit widening rate W has been determined at different temperatures. The graphs obtained by plotting log of etch pit widening rate W against the reciprocal of temperature are shown in Figure 3. The slope of the graph gives a definite value of activation energy to be 0.28 eV. This is about half the value for NaCl, comparison of this value with the reported values of NaCl indicates that the dissolution is instantaneous at high temperatures. This conclusion is confirmed by physical observation of etch pits taken at high temperatures (SANGWAL, SUTARIA).

The formation of etch pits is due to the production of nucleated pits on the defects of the crystal structure, and the subsequent movement of the resultant steps in the tangential direction. If the pits are formed on dislocations, their depth is determined by the nucleation rate along the dislocation line. The square form and the orientation of the edges of the base of the etch pits in water parallel to $\langle 100 \rangle$ is due to the fact that in the $\langle 100 \rangle$ direction the rate of movement of the elementary dissolution steps is a minimum (SANGWAL et. al.).

The author is thankful to Dr. V. Hari Babu and the Head, Department of Physics, Osmania University, for their kind encouragement to undertake the present investigation.

References

BHAT, H. L., NARAYANAN, P. S.: J. Mat. Sci. **10**, 2007 (1975)
KOSTIN, N. F., LUBENETS, S. V., ALEKSANDROV, K. S.: Sov. Phys. Cryst. **6** (5) 588 (1963)
PATEL, A. R., SINGH, R. P.: J. Cryst. Growth **5**, 70 (1969)
ROZHANSKI, V. N., PARVOVA, E. V., SQEPNOVA, V. M., PREDVODITELEV, A. A.: Sov. Phys. Cryst. **6** (5) 564 (1962)
SANGWAL, K., SUTARIA, J. N.: J. Mat. Sci. **11**, 2271 (1976)
SANGWAL, K., URUSOVSKAYA, A. A., SMIRNOVA, A. E.: Ind. J. Pure and Appl. Physics **16**, 501 (1978)
SOMAIAH, K., HARI BABU, V.: J. Cryst. Growth **46**, 711 (1979)
SOMAIAH, K., HARI BABU, V.: Cryst. Res. Technol. **18** (9) 1101 (1983)

(Received February 23, 1987)

Author's address:

Dr. K. SOMAIAH
Department of Physics, Osmania University
Hyderabad-500007
India

Cryst. Res. Technol.	22	1987	9	K145–K147

G. BEDNARZ, M. J. KRASINSKI

Institute of Physics, Technical University of Łódź, Poland

Visualization of the Boundary Layer on the Growing Ice

Introduction

Recently, the investigations of the interfacial layer between crystal and solution have been in progress. In a water-ice system interferometric techniques can be used for studying density changes of solution near growing ice (TIBY et al.). In the paper we present results of investigations of the water-ice interface formed from pure water by Mach-Zehnder interferometer. The interferometer has been used to obtain an experimental evidence for existence of the transitional layer between bulk water and ice.

Experimental

The flat experiment cell with the double side-walls made from plexiglass was used (Fig. 1). The cell was several millimeter thick. The bottom of the cell was cooled by liquid nitrogen vapour, which enabled ice to grow vertically to the bottom of the cell. The temperature inside the cell was measured by a thermocouple and recorded in a function of time by a chart recorder. The thermojunction was placed somewhat below the water surface. The water-ice interface was led vertically to the direction of the ice growth by He–Ne laser. The interference fringes were recorded by a camera.

For measurements the distilled, non-degased and degased water were used. The degased water was prepared by boiling at a low pressure. Samples were cooled at a rate of 1 – 2 K/min. The rate of the ice growth was estimated to be 20 – 25 mm/min at the beginning of freezing and 10 – 13 mm/min at the end.

Refractive index change Δn was evaluated basing on a shift of interference fringes (Fig. 2) by formula

$$\Delta n = (\Delta x \cdot \lambda)/(d \cdot \Delta l)\,, \tag{1}$$

where d is the thickness of the cell (3 mm), Δx is shift of the interference fringes, λ is the wavelength of the laser light (638 nm), Δl is the distance between the consecutive inter-

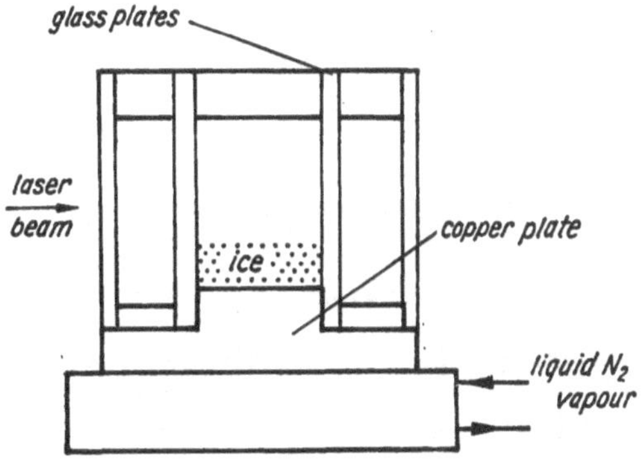

Fig. 1. Schematic drawing of the experiment cell

ference fringes. The sensitivity of our setup defined as the minimum detectable refractive index change was about $3 \cdot 10^{-5}$. It gives the experimental error below 15%.

Results

The shift of interference fringes in the water adjacent to the advancing ice front was detected (Fig. 2a). Having examined diffractive properties of the setup we found out that the shift is caused by an increase of the refractive index of water in the vicinity of the water-ice interface. For the non-degased water the thickness of the boundary layer was several tens of microns and the maximum change of the refractive index of water was about $4 \cdot 10^{-4}$. The boundary between the ice and the water was sharp and well visible. The ice below the boundary was non-transparent. There were visible many air bubbles captured by growing ice. When a frozen sample was melted and cooled again the interference pattern was sometimes distinctly different (Fig. 2b). The interference fringes were blurred and declined gradually in the boundary region. The fringe shift increased considerably. The boundary layer thickness could roughly be estimated to be about 0.5 mm. The change of the refractive index was in the range $2-3 \cdot 10^{-3}$. In such cases the ice was more transparent than after first cooling. Thus the abrupt increase in the refractive index may have been caused by an increase in the transparency of the ice near the ice-water boundary due to liberating of air from freezing water. We conducted the comparison measurements for the degased water. Despite an increase in the transparency of the ice, the interference patterns were like in the Figure 2a.

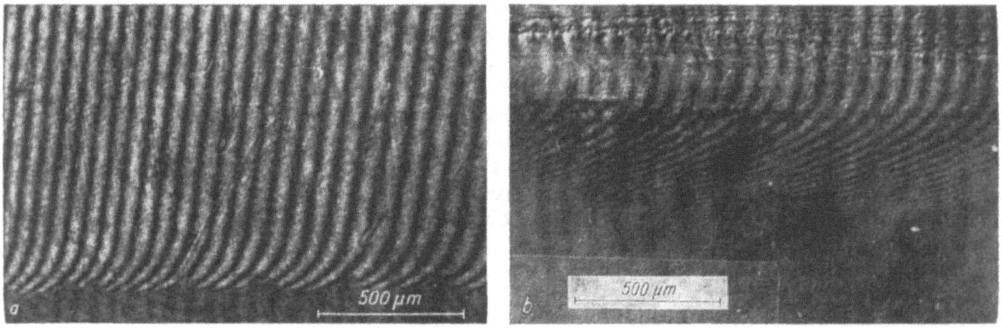

Fig. 2. The interference pattern (a) for the non-degased water and (b) observed in some cases for the second freezing of the non-degased water

The observations listed above relate to the case of the advancing ice front. When freezing of a sample was hindered, the fringe shift declined. After starting the freezing, the fringe shift arose again.

During measurements the temperature of water under a water surface was recorded. When ice became apparent on the bottom of the cell the temperature dropped below 4 °C. The temperature of water was in the range 1—2 °C in the distance of 0.5—1 mm from the ice front and increased with an increase in the rate of freezing.

Discussion

The fringes shift may be attributed to an increase of the density of water in the layer adjacent to the water-ice interface. From literature data (BALLARD et al.; GMELIN's Handbuch der Anorganischen Chemie) it follows that the refractive index of water increases by about $0.5 \cdot 10^{-4}$ when temperature decreases from 5 °C to 0 °C. Comparing this quantity with our results and taking into account the estimation of the temperature changes within the sample one can say that the fringes' shift is not caused by the temperature variations of the density of water. DROST-HANSEN and also ROSENTHAL postulated the existence of the boundary layer between water and ice to explain a nature of electrical potential occurring during freezing of dilute aqueous solutions. According to them, the boundary layer may involve a transitional, disordered structure between the possessing local order water and the crystal lattice of ice. The disordering extends into liquid and diminishes in a function of the distance from the interface. Thus, the increase in the water density may be caused by the increase in the disordering of water molecules in the boundary layer during freezing. The layer may be too thin to be detected by our setup in the stationary case.

We cannot explain the origin of the interference pictures like in Figure 2b. We suppose we detected a greater part of the boundary layer in these cases. It was probably due to profitable conditions of observations of the layer. Among the factors affecting this conditions one can mention the amount of gases dissolved in water and the rate of freezing. On the other hand a preliminary freezing and melting of a sample may directly influence the interference picture of the boundary layer during the second freezing. This phenomenon requires further investigations.

This work was carried out under Research Project CPBP 01.20.

References

BALLARD, S. S., McCARTHY, K. A., BROUWER, W.: in: American Institute of Physics Handbook, New York 1957

DROST-HANSEN, W.: J. Coll. Int. Sci. 25, 131 (1967)

GMELIN's Handbuch der anorganischen Chemie, Auflage 8, vol. 3, Weinheim 1963

ROSENTHAL, O. M.: Zh. Strua. Khimii 9, 777 (1968) (in Russian)

TIBY, C., BEHRET, M., ERDMANN, M., LANGBEIN, D.: in: Report 1984, Battelle Institute e. V. Frankfurt (in German)

(Received March 20, 1987)

Authors' address:

G. BEDNARZ, M. J. KRASINSKI
Institute of Physics, Technical University of Łódź
ul. Wolczanska 219
93-005 Łódź, Poland

Cryst. Res. Technol.	22	1987	9	K148–K150

D. LINGA REDDY, M. SURYANARAYANA

Post Graduate Centre (Osmania University) Godavari Khani, India, and
Department of Physics, Osmania University, Hyderabad, India

Elastic Constants C_{11} and C_{44} of Mixed Crystals KCl–KBr

Introduction

Ultrasonic work in the determination of second order elastic constants (SOEC) of pure crystals has been reported in literature (KRISHNA MURTHY, SURYANARAYANA; TRUELL et al.). But the earlier work on the elastic constants of mixed crystals is scanty. The work on the elastic constants of solid solutions of KCl–KBr by SLAGLE and MCKINSTRY by phase comparison technique as a function of concentration of one of the components and of temperature is worth mentioning. CAIN also found the elastic constants and their pressure derivatives of KCl–KBr mixed crystals. SHANKAR and JAIN calculated theoretically the SOEC of KCl–KBr mixed crystals. KATAOKA and YAMADA have given the SOEC of different portions of Br in KCl. The work of SRIVASTAVA and of BASU et al. should also be mentioned in this connection. VEERESHAM et al. determined thermoluminescence and optical absorption in KCl–KBr mixed crystals. SMAKULA and MAYNARD worked on the hardness as a function of concentration of KBr in KCl. KOPKINS et al., and SUBBA RAO and HARI BABU worked on some mechanical properties of KCl–KBr mixed crystals. KAMIYOSHI and NIGARA worked on the dielectric constants of KCl–KBr mixed crystals. LINGA REDDY and SURYANARAYANA carried out the work on elastic constants of alkali halide mixed crystals.

Hence it is desirable to determine experimentally the SOEC of KCl–KBr mixed crystals as a function of composition with a very accurate pulse superposition method of MCSKIMIN now available commercially and of sufficient accuracy.

Experimental

The ultrasonic pulse echo interferometer based on MCSKIMIN's pulse superposition technique is supplied by M/s. Systems Dimensions, Bangalore, on the technical know-how from SRINIVASAN et al. of the Indian Institute of Science, Bangalore, India. Some modifications have been done in this interferometer.

In the pulse superposition method now used, a transducer of piezoelectric quartz (x-cut for longitudinal wave and y-cut for transverse waves) having fundamental frequency of 10 MHz is attached to the specimen through a bond of salol (phenyl salicilate of melting point 42 °C). The ultrasonic pulsed energy is reflected from the opposite plane parallel face of the specimen and received by the same transducer. The echo pattern received by the transducer after suitable amplification is received by the cathode ray osciloscope (CRO) and the time interval T, between successive echoes, is measured by a six digit frequency/timer counter of 30 MHz with an accuracy of ± 0.1 Hz. The velocity v of the wave is $v = 2l/T$, where l is the length of the specimen an $C_{ij} = \varrho v^2$, where C_{ij} is C_{11} if v is longitudinal and is C_{44} if v is transverse. ϱ is the density of the specimen. The above formula is valid for ⟨100⟩ cut of a cubic crystal.

An overall precision of 1 part in 10^5 in the evaluation of T is obtained. The accuracy in the velocity measurements depend on the accuracy with which the sample thickness is measured and a screw gauge of least count 0.001 cm is now used and so the accuracy of the velocity measurements is of the order of ± 0.05 m/s.

Again, the accuracy in the evaluation of the elastic constants depends on the accuracy with which the density of the specimen is measured. Since the density of the specimen is here determined by the principle of Archimedes and the values of density are correct to ± 0.005 g/cm³ and the values of C_{11} and C_{44} are correct to $\pm 0.005 \cdot 10^{11}$ dynes/cm². The elastic constant C_{12} is not determined as the samples are very small in size.

The crystals of pure KCl, pure KBr and various compositions of KCl–KBr mixed crystals were grown in air by Kyropoulos technique and the KCl and KBr are of BDH analar grade salts. All the crystals were grown and annealed at 650 °C under identical conditions.

Results

The elastic constants C_{11} and C_{44} are determined for ⟨100⟩ cut and tabulated in Table 1 for mixed solid solutions KCl–KBr near 36 °C for C_{11} and near 30 °C for C_{44} at three concentrations 45 mol%, 61 mol%, and 72 mol% KBr in KCl. In Table 1 the authors have given the values of C_{11} and C_{44} of pure KCl, pure KBr and three compositions of the solid solutions of KCl–KBr mixed systems and compared the values with those obtained by Slagle and McKinstry. Unfortunately Slagle and McKinstry gave the C_{11} and C_{44} values for concentrations not taken by the authors, namely, 45 mol%, 61 mol% and 72 mol% of KBr in KCl. Hence the authors calculated C_{11} and C_{44} values at the required concentrations and incidentally also C_{11} is calculated at 35 °C and C_{44} at 30 °C. For this, C_{ij} of the formula (Slagle, McKinstry),

$$C_{ij} = C_{ij}^0 + \alpha_{ij} T$$

are calculated at 35 °C for C_{11} and 30 °C for C_{44}. Here C_{ij}^0 is the elastic constant at zero degree Kelvin and α_{ij} is the temperature coefficient of elastic constant and T is the absolute temperature. The formula is valid for temperatures above Debye characteristic temperatures. Since Debye temperature of KCl is 236.3 K and KBr is 172.8 K use can be made of the formula (Slagle, McKinstry). C_{ij}^0 and α_{11} are given and α_{44} calculated. The authors' values of C_{44} (Tab. 1) are much coinciding with the values of Slagle and McKinstry, whereas in C_{11} they are slightly differing. But as

Table 1
The elastic constants C_{11} and C_{44} in units of 10^{11} dynes/cm² in solid solutions of KCl–KBr

Crystal	C_{11}			C_{44}		
	temp. °C	authors' value	Slagle, McKinstry*)	temp. °C	authors' value	Slagle, McKinstry*)
0.0 (pure KCl)	35	4.10	4.04	30	0.59	0.63
45% KBr	35	3.66	3.68	30	0.57	0.57
61% KBr	35	2.96	3.60	30	0.51	0.55
72% KBr	35	2.89	3.55	30	0.50	0.54
100% KBr (pure)	35	2.80	3.44	30	0.44	0.50

*) Recalculated by the authors (see text).

a conclusion the values of C_{11} and C_{44} are decreasing linearly from pure KCl to 45% KBr, 61% KBr, 72% KBr, and 100% KBr (pure KBr) in KCl.

The authors are thankful to Dr. Katta Narasimha Reddy for giving samples. One of the authors (D.L.R.) is grateful to Incharge and Staff of P. G. Centre, Godavarikhani, for their encouragement and help. Financial assistance from the UGC, New Delhi, in the form of a minor research project to D.L.R. is thankfully acknowledged.

References

BASU, C., KANRAR, A., GHOSH, U. S.: phys. stat. solidi (b) **114**, 221 (1982)
CAIN, L. S.: J. Phys. Chem. Solids **37**, 1178 (1976)
HOPKINS, J. R., MILLER, J. A., MARTIN, J. J.: phys. stat. solidi (a) **19**, 591 (1973)
KAMIYOSHI, K., NIGARA, Y.: phys. stat. solidi (a) **6**, 223 (1971)
KATAOKA, T., YAMADA, T.: Jpn. J. Appl. Physics **16**, 1119 (1977)
KRISHNAMURTHY, M., SURYANARAYANA, M.: Elastic properties of crystals as studied by ultrasonics, New York 1969, p. 487
LINGA REDDY, D., SURYANARAYANA, M.: Acoustic letters **9**, 175 (1986); Cryst. Res. Technol. **22**, K88 (1987)
MCSKIMIN, H. J.: J. Acoust. Soc. Amer. **33**, 12 (1961)
SHANKAR, J., JAIN, G. D.: Phys. Rev. B **27**, 2515 (1983)
SLAGLE, O. D., MCKINSTRY, H. A.: J. Appl. Physics **38**, 437, 446 (1967)
SMAKULA, A., MAYNARD, M., REPUCCI, A.: J. Appl. Physics **33**, 453 (1962)
SRINIVASAN, K. R., GOPAL, E. S. R.: Pramana **6**, 19 (1976)
SRIVASTAVA, U. C.: J. Appl. Physics **51**, 1510 (1980)
SUBBA RAO, U. V., HARIBABU, V.: PRAMANA **11**, 149 (1978)
TRUELL, R., ELBAUM, C., CHICK, B. B.: Ultrasonic methods in solid state physics, New York 1969, p. 370
VEERESHAM, P., NARSIMHAREDDY, K., SUBBARAO, U. V., HARI BABU, V.: Cryst. Res. Technol. **18**, 1585 (1983)

(Received March 23, 1987)

Authors' addresses:

Dr. D. LINGA REDDY
Post Graduate Centre (Osmania University)
Godavari Khani-505209, India

Prof. M. SURYANARAYANA
Department of Physics, Osmania University
Hyderabad-500007., India

Cryst. Res. Technol.	22	1987	9	K151–K153

V. S. Yadav, S. K. Chaudhary, V. K. Jain

Department of Physics, M. D. University, Rohtak, India, and
Department of Physics, University College, Rohtak, India

Purification and Growth of Ferroelectric Sodium Nitrite Single Crystals Using Zone-refining Technique

Introduction

Hitherto some experimental work on the growth of ferroelectric sodium nitrite ($NaNO_2$) single crystals has been carried out by slow evaporation as well as by melt (Beznosikov). However, little or no work have been done on the purification of the material. In this paper we report on a zone-refining system fabricated in our laboratory, purification and growth of single crystals of $NaNO_2$. The crystal structure of $NaNO_2$ is orthorhombic with $a_0 = 3.569$ Å; $b_0 = 5.563$ Å; $C_0 = 5.384$ Å (Wyckoff).

Experimental

For the purification and growth of single crystals of $NaNO_2$ a zone-refining system was set up in the laboratory (Fig. 1). Since the melting point of $NaNO_2$ is 271 °C, the main chamber consisted of pyrex glass. It has a diameter of nearly 3 cm and length of nearly 40 cm. A vacuum-tight contact was ensured at this end by sealing the end with picein wax. The brass tube ended on the other side into a flat metallic flap which had a circular groove in it for holding on O-ring seal. This flat face was screwed into other similar brass tube, which had an exactly matching flat surface with a groove and nitrogen gas inlet tube. The open end of the tube was dipped into a beaker of water from where the nitrogen gas could escape slowly. To save time, the separate heat zone was provided along the length

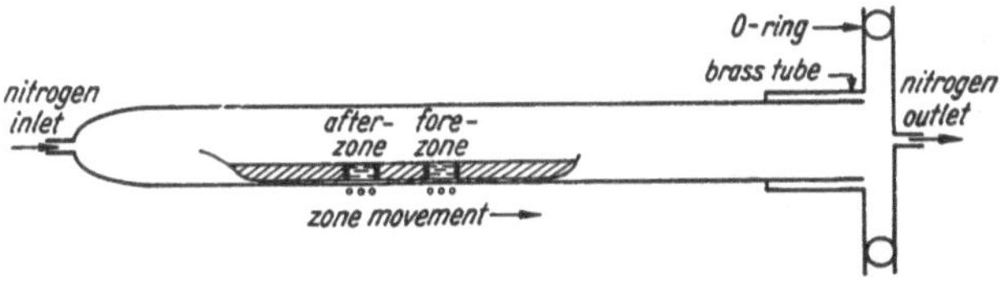

Fig. 1. Line diagram of a zone-refining system

of the chamber by coaxially mounted, wire-wound semi-circular resistance heaters. The length of each zone was kept nearly 1 cm. The heaters were mounted on a trolley which could move on a pair of guide rails. Necessary steps were taken to reduce the vibrations in the system to a minimum. For movement of heater the trolley was connected to a motor-driven slow speed gear and pulley arrangement. The speed of heaters could be adjusted by using pulleys of different diameters. The zone length could be controlled by adjusting the voltage fed to the heaters as well as by adjusting the height of the heaters.

Results and discussion

Initially the pyrex glass boats were employed for the growth of $NaNO_2$ crystals as melting point of the latter being 271 °C. After just one zone passage the boat was found to break in each attempt. It was attributed to the anisotropic thermal properties of $NaNO_2$ which on cooling contracts along *a*- and *c*-axis and expands along *b*-axis (BEZNOSIKOV).

Therefore the crystals were grown in teflon boats. Powdered $NaNO_2$ (purity 98%) (supplied by E. Merck, FRG) was put into the boat which was then introduced into the chamber such that it lays in the central region of chamber. The nitrogen gas was slowly admitted into the system. One of the heaters was placed at the tempered end of the boat and the other at a distance of nearly 2 cm from it. Necessary power was fed to the heater to make the sodium nitrite salt melt. The speed of movement of heaters was found by hit and trial and a most suitable value was found to be 2 cm/h. After each zone passage the heaters were returned to their initial position by manual operation, whence the process of zone refining could be started once again. In the final zone passage the height of the afterheater was suitably reduced to permit a slow cooling of the crystal.

Figure 2 shows the temperature profile for the growth of best possible crystals.

At the end of first zone passage, the initial part of the material appeared fairely transparent. A good quality single crystals could be grown after ten passages. Figure 3 shows one of the crystals grown by this method. To verify that the ingot had grown as single crystal Laue photographs of the ingot were taken from its different regions. The initial part of the crystal was found to be polycrystalline. This was expected, because in the beginning of crystal growth many different nuclei start forming and eventually the nucleus having the preferred orientation for growth crowds out of others with less favourable orientation (CHALMERS). The final portion of the crystal was also found to be polycrystalline, which could be attributed to the accumulation of impurities at that end during zone-refining. However, the intermediate section of the crystal measuring nearly three centimeters in length was found to be a single crystal as all the Laue photographs taken from its different regions showed well-defined discrete reflections of nearly the same shape and size. There is no doubt that transparency of the crystal is one of the indices of purification.

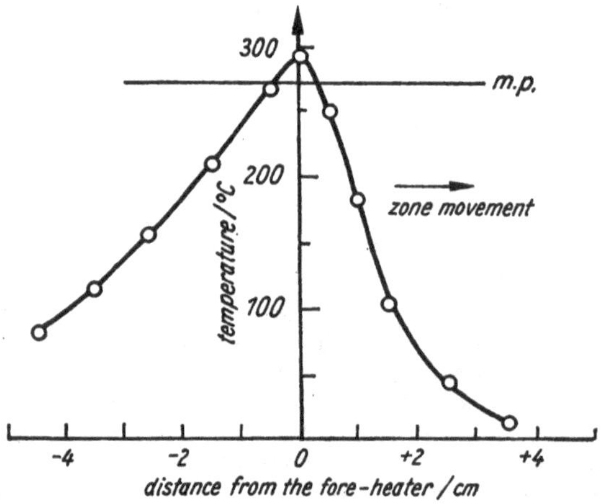

Fig. 2. Temperature profile for the growth of $NaNO_2$ single crystals

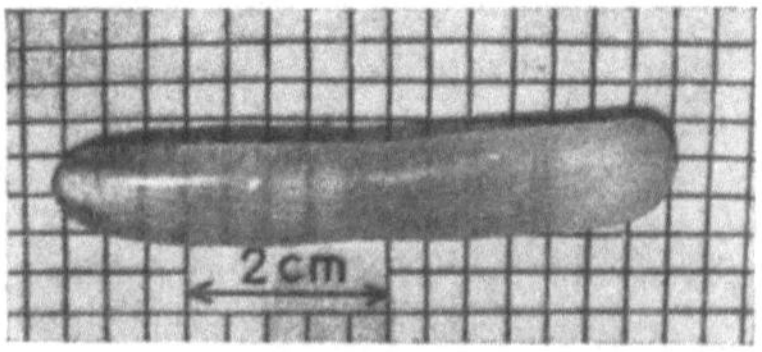

Fig. 3. Photograph showing the crystal of $NaNO_2$ grown by a zone-refining system

But to be doubly sure that purification is being done by this system, the melting points of samples taken from three regions of the grown ingot, i.e. initial region (pure), middle region and the final region (impure) were determined at 74.58 cm of mercury and were found to be

Initial region: 270–271 °C

Middle region: 269.5–270.5 °C

Final region: 267–269 °C

which is an expected result as the impurities lower down the melting point.

This work was supported by C.S.I.R., and U.G.C., New Delhi.

References

BEZNOSIKOV, B.V.: Soviet Phys. Cryst. (USA) **10**, 214 (1965)
CHALMERS, B.: Canadian J. Phys. **31**, 132 (1953)
WYCKOFF, R. W. G.: Crystal structures, II. Ed. Vol. 2, New York 1964, p. 302

(Received March 25, 1987)

Authors' addresses:

V. S. YADAV, Dr. V. K. JAIN
Department of Physics, M. D. University
Rohtak-124001, India

Dr. S. K. CHAUDHARY
Department of Physics, University College
Rohtak-124001, India

Cryst. Res. Technol.	22	1987	9	K154–K157

LI JIANMIN, PAN GUOQIANG, FANG JIFA

Department of Modern Chemistry, China University of Science and Technology, Hefei, Anhui, P.R. China

Measurement and Explanation of the d-d-Transition Spectra of CaM(II)-EDTA Crystal. 1. Ca[Co(EDTA)] · 5 H_2O

Introduction

Ethylene diamine tetra acetic acid (EDTA) and its double salt are ligands of the metal ions. They are widely used in the analytic chemistry. Therefore workers are very interested in them. Many researchers have determined the crystal structure of the double salt CaM(II)-EDTA. (YUJIDAEV; SMITH et al.; STEPHANS; NESTEROVA 1982, 1984.) Their d–d transition spectra are not extensively studied so far. However, the measurement and reasonable explanation of the electronic absorption spectrum is important to study the electronic structure of the coordination compounds. Recently we have measured a series of the d–d transition spectra of CaM(II)-EDTA and explained them reasonably with the theory proposed by us. We will report them respectively. This paper is the first one.

Experiment

The crystal used in this investigation was prepared by mixing aquesous solution of EDTA, calcium chloride and cobalt chloride with a molar ratio 1:1:1. The solution were slowly evaporated at 60 °C, then were placed several days at room temperature. The deep purple crystal were obtained.

The electronic absorption spectrum of Ca[Co(EDTA)] · 5 H_2O crystal was measured in the region of 4000–30000 cm^{-1} with the electronic absorption spectrograph for a small crystal which was grown by the Institute of Geochemistry, Academia Sinica. The results measured are shown in Figure 1.

Theoretical explanation

We calculated the d–d transition spectrum of Ca[Co(EDTA)] · 5 H_2O crystal on the basis of the ligand field theory. It can be seen from the data of the crystal structure that the symmetry of Ca[Co(EDTA)] · 5 H_2O crystal is very low. We chose O(5)–Co–O(1) as symmetry axis (see Fig. 2). The approximate symmetry of Co(II) ions is C_s. The energy matrix of d^7 configuration in C_s crystal field (weak field scheme) is:

$${}^4A'({}^4A_{2g}, {}^4F): \qquad 28b_3 + 4b_5$$

$$\begin{array}{ccc} {}^4A'({}^4T_{2g}\zeta, {}^4F) & {}^4A'({}^4T_{1g}z, {}^4F) & {}^4A'({}^4T_{1g}z, {}^4P) \end{array}$$

$$\begin{bmatrix} 28b_3 - 4b_5 & -\sqrt{\frac{3}{5}}\left(10b_2 - \frac{25}{2}b_4\right) & -2\sqrt{\frac{3}{5}}\left(10b_2 - \frac{25}{2}b_4\right) \\ & 4b_1 - 24b_2 & -12b_1 + 16b_3 \\ & & -14b_1 + 15B \end{bmatrix}$$

$$
\begin{array}{lll}
{}^4A'({}^4T_{2g}\eta, {}^4F) & {}^4A'({}^4T_{1g}y, {}^4F) & {}^4A'({}^4T_{1g}y, {}^4P) \\
\left[\begin{array}{l} -7b_3 + 35b_4 + 3b_5 \\ \\ \\ \\ \\ \end{array}\right. & \begin{array}{l} \sqrt{\tfrac{3}{5}}(5b_1 - 5b_2 + 5b_3 + 5b_4 - b_5) \\ \\ -2b_1 - 6b_2 - 9b_3 - 15b_4 - 3b_5 \\ \\ \\ \end{array} & \left.\begin{array}{l} \\ 2\sqrt{\tfrac{3}{5}}(5b_1 + 5b_2 + 5b_3 - 5b_4 - b_5) \\ \\ 6b_1 + 18b_2 + 6b_3 + 10b_4 + 2b_5 \\ 7b_1 + 21b_2 + 15B \end{array}\right]
\end{array}
$$

$$
\begin{array}{lll}
{}^4A'({}^4T_{2g}\xi, {}^4F) & {}^4A'({}^4T_{1g}\chi, {}^4F) & {}^4A'({}^4T_{1g}\chi, {}^4P) \\
\left[\begin{array}{l} -7b_3 - 35b_4 + 3b_5 \\ \\ \\ \\ \\ \end{array}\right. & \begin{array}{l} \sqrt{\tfrac{3}{5}}(5b_1 + 5b_2 + 5b_3 - 5b_4 - b_5 \\ \\ -2b_1 + 6b_2 - 9b_3 + 15b_4 - 3b_5 \\ \\ \\ \end{array} & \left.\begin{array}{l} \\ 2\sqrt{\tfrac{3}{5}}(5b_1 - 5b_2 + 5b_3 + 5b_4 - b_5) \\ \\ 6b_1 - b_2 + 6b_3 - 10b_4 + 2b_5 \\ 7b_1 - 21b_2 + 15B \end{array}\right]
\end{array} \tag{1}
$$

where the coefficients b_1, b_2, b_3, b_4, b_5 are as follows:

$$b_1 = \frac{1}{70}\bar{N}^2K_1P^{(2)}\langle r^2\rangle\left[\frac{4}{R_{11}} - \left(\frac{1}{R_1^3} + \frac{1}{R_2^3} + \frac{1}{R_3^3} + \frac{1}{R_4^3} + \frac{1}{R_5^3}\right)\right]$$

$$b_2 = \frac{1}{70}\bar{N}^2K_1P^{(2)}\langle r^2\rangle\left[\frac{1}{R_1^3} + \frac{\cos 2\Phi_1}{R_2^3} + \frac{1}{R_3^3} + \frac{\cos 2\Phi_2}{R_4^3} + \frac{\cos 2\Phi_3}{R_5^3}\right]$$

$$b_3 = \frac{1}{672}\bar{N}^2K_1P^{(4)}\langle r^4\rangle\left[\frac{16}{R_{11}^5} + 3\left(\frac{1}{R_1^5} + \frac{1}{R_2^5} + \frac{1}{R_3^5} + \frac{1}{R_4^5} + \frac{1}{R_5^5}\right)\right]$$

$$b_4 = -\frac{1}{168}\bar{N}^2K_1P^{(4)}\langle r^4\rangle\left[\frac{1}{R_1^5} + \frac{\cos 2\Phi_1}{R_2^5} + \frac{1}{R_3^5} + \frac{\cos 2\Phi_2}{R_4^5} + \frac{\cos 2\Phi_3}{R_5^5}\right]$$

$$b_5 = \frac{35}{168}\bar{N}^2K_1P^{(4)}\langle r^4\rangle\left[\frac{1}{R_1^5} + \frac{\cos 4\Phi_1}{R_2^5} + \frac{1}{R_3^5} + \frac{\cos 4\Phi_2}{R_4^5} + \frac{\cos 4\Phi_3}{R_5^5}\right]. \tag{2}$$

The coefficients b_1, b_2, b_3, b_4, b_5, Racah parameters B, C and ζ_{3d} are calculated with the average environment rule (ROSENBERG et al.) because $Ca[Co(EDTA)] \cdot 5\,H_2O$ is mixedly coordinated by oxygen and nitrogen.

According to the salling radial wave function theory of non-free ion, the double-ζ radial wave function of Co(II) can be written as: (ZHANG YONGFENG)

$$R_{3d}(r, \Omega) = C^{-1/2}\left[0.509614\sqrt{\frac{(2\zeta_1)}{6!}}\,r^2\,e^{-\zeta_1 r} + 0.65\sqrt{\frac{(2\zeta_2)}{6!}}\,r^2\,e^{-\zeta_2 r}\right], \tag{3}$$

where

$$\zeta_1 = 5.9068(1 - 0.23500\Omega - 0.22160\Omega^2)$$

$$\zeta_2 = 2.3250(1 - 1.52803\Omega + 0.396771\Omega^2)$$

$$C = 1 - 0.663498\left[0.479689 - \left(\frac{2\sqrt{\zeta_1\zeta_2}}{\zeta_1 + \zeta_2}\right)^7\right]. \tag{4}$$

From the data of the crystal structure we obtain $R_1 = 2.2154$ Å, $R_2 = 2.1994$ Å, $R_3 = 0.0595$ Å, $R_4 = 2.1955$ Å, $R_5 = 2.7435$ Å, $R_{11} = 2.1075$ Å, $\Phi_1 = 79.2°$, Φ_2

= 88.47°, Φ_3 = 144.6°. The non-free radial wave function of Co(II) is determined by formula (3), (4) and the crystal structure. Thus some crystal field parameters can be calculated by it. The results calculated are listed in Table 1.

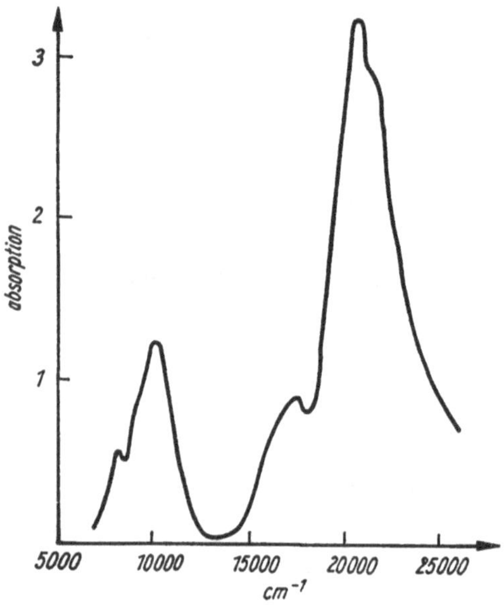

Fig. 1. The electronic absorption spectrum of Ca[Co(EDTA)] · 5 H_2O

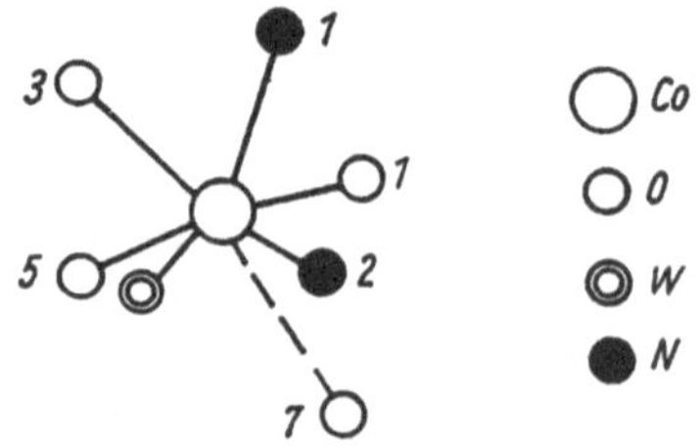

Fig. 2. The crystal structure of Ca[Co(EDTA)] · 5 H_2O

Table 1

The crystal field parameters of Ca[Co(EDTA)] · 5 H_2O

parameter	ligand	
	O	N
Ω (Hartree)	0.14419	0.16459
μ (Debye)	0.961	1.148
$\bar{R}$ (Å)	2.117	2.207
q	1.06	0.80
q'	6	2
t	0.017056	0.054147
$P^{(2)}$	1.43384	1.307989
$P^{(4)}$	1.69386	1.695151
ζ_1	5.679434	5.642880
ζ_2	1.831917	1.765267
a_1	0.534042	0.537707
a_2	0.681176	0.685832
$\langle r^2 \rangle$ (a.u.)	2.3084	2.4857
$\langle r^4 \rangle$ (a.u.)	13.4613	15.7505
B (cm^{-1})	923.953	890.816
C (cm^{-1})	3278.616	3151.866
ζ_{3d} (cm^{-1})	446.66	433.86
$\bar{N}^2$	0.980	0.977

Table 2
The electronic absorption spectrum of $Ca[Co(EDTA)] \cdot 5\,H_2O$

Assignment	calc.	obs.	assignment	calc.	obs.
$E'(^4T_{1g}z, {}^4F)$	0		$^4A'(^4T_{2g}z, {}^4F)$	8550	9100
$^4A'(^4T_{1g}y, {}^4F)$	1010		$^4A'(^4A_{2g}', {}^4F)$	15220	16550
$^4A'(^4T_{1g}z, {}^4F)$	1920		$^4A'(^4T_{1g}x, {}^4F)$	19460	
$^4A'(^4T_{2g}y, {}^4F)$	6750	7150	$^4A'(^4T_{1g}z, {}^4P)$	20410	19650
$^4A'(^4T_{2g}x, {}^4F)$	7700	8250	$^4A'(^4T_{1g}y, {}^4P)$	21320	20700

Using the data of Table 1 and formula (1), (2), we calculated the transition energy level of the $Ca[Co(EDTA)] \cdot 5\,H_2O$ crystal field. The energy of spin-orbit coupling calculated in the O_h field approximation is added into the ground state because it has more influence on the energy level of crystal field. The results calculated are listed in Table 2. They are nearly in agreement with the experimental value. The errors originate mainly from the central potential field approximation because of the low symmetry of the $Ca[Co(EDTA)] \cdot 5\,H_2O$ crystal.

References

NESTEROVA, Ya. M.: Coord. Chem. 8, 994 (1982); **10**, 129 (1984) (in Russian)
ROSENBERG, R. C., ROOT, C. A., GRONY, H. B.: J. Amer. Chem. Soc. **97**, 21 (1975)
SMITH, G. S., HOARD, L. J.: J. Amer. Chem. Soc., **81** (3) 556 (1959)
STEPHANS, S. F.: J. Chem. Soc. A(11) 1723 (1969)
YUJIDAEV, A. I.: J. Struct. Chem. **13**, 244 (1972) (in Russian)
ZHANG YONGFENG: M. S. thesis, China University of Science and Technology (1981)

(Received March 27, 1987)

Authors' address:

LI JIANMIN, PAN GUOQIANG, FANG JIFA
China University of Science and Technology
Hefei, Anhui
P.R. China

Book Review

W. KARCHER, R. J. FORDHAM, J. J. DUBOIS, P. G. J. M. GLAUDE, J. A. M. LIGTHART (Hrsg.)

Spectral Atlas of Polycyclic Aromatic Compounds

D. Reidel Publishing Company, Dordrecht/Boston/Lancaster 1985, 818 S., Preis: 94 US $/ Dfl. 280.00/ £77.75. SBN 90-277-1652-8

Polycyclische aromatische Kohlenwasserstoffe und ihre Derivate einschließlich einiger heteroaromatischer Verbindungen, die beispielsweise bei der unvollständigen Verbrennung organischer Verbindungen ständig gebildet werden, sind in der Umweltanalytik von großer Bedeutung, da viele von ihnen kanzerogene und mutagene Eigenschaften aufweisen.

Die analytische Bestimmung dieser polycyclischen Verbindungen im Spurenbereich wird meist dadurch erschwert, daß keine Vergleichsdaten zur Verfügung stehen. Es ist deshalb sehr begrüßenswert, wenn von den Autoren eine Sammlung verläßlicher Spektraldaten für 43 ausgewählte polycyclische aromatische Verbindungen präsentiert wird, die in einer Reinheit von 99% und besser vorlagen. Von diesen Verbindungen wurden von ihnen in guter Qualität UV-, IR-, ^{1}H-NMR-, ^{13}C-NMR-, Massen- und Fluoreszenz- sowie Tieftemperatur-Fluoreszenz- bzw. Phosphoreszenzspektren vermessen, wobei teilweise unterschiedliche apparative Voraussetzungen — z. B. Messung der ^{1}H-NMR-Spektren mit einem 80 MHz- oder einem 300 MHz-Spektrometer — vorlagen. Die Meßbedingungen sowie die verwendeten Geräte sind jeweils angegeben. Die Spektraldaten sind in übersichtlicher Weise sowohl in Form der Spektren als auch tabellarisch dargestellt. Soweit bekannt wurde eine Signalzuordnung vorgenommen. Ergänzt werden diese Spektraldaten durch Tabellen über das Vorkommen dieser Verbindungen (in fossilen Brennstoffen, in der Umwelt, in Nahrungsmitteln) sowie über ihre biologische Aktivität (Mutagenität, Kanzerogenität, Toxizität) und andere Eigenschaften.

Dem Rezensenten ist nicht ganz klar, nach welchen Kriterien die Auswahl der vermessenen Substanzen erfolgte. Er vermißt darin Spektraldaten über Derivate des Dibenzodioxins, Dibenzofurans, Diphenyls, die eine wesentliche Rolle in der Umweltanalytik spielen und deren Aufnahme die Attraktivität dieses Atlasses zweifellos weiter gesteigert hätte.

Diese Bemerkungen können jedoch die Nützlichkeit des vorgelegten Spektrenatlasses in keiner Weise schmälern. Für jeden, der Umweltanalytik betreibt, ist diese Sammlung von außerordentlichem Wert.

R. BORSDORF

Book Review

B. Wilshire, R. W. Evans (Hrsg.)

Creep and Fracture of Engineering Materials and Structures

Proceedings of the Third International Conference, Swansea, 5.—10. April 1987

The Institute of Metals London, 1987, 1043 Seiten, zahlreiche Abb., UK £ 58.00, US $ 110.00
ISBN 0-904357-99-6

Die zunehmende Bedeutung der Werkstoffmechanik für die Verbesserung der Zuverlässigkeit und technischen Sicherheit von Maschinen und Anlagen hat auch in der Durchführung internationaler Konferenzen zu den verschiedenen Aspekten der Werkstoffschädigung, insbesondere zum Bruchverhalten von Werkstoffen und Konstruktionen, ihren Ausdruck gefunden. Der vorliegende Tagungsband umfaßt Originalarbeiten zum Kriechen bzw. Bruch bei höheren Temperaturen, die in folgenden Sektionen angeordnet sind:

— Mechanismen des Kriechens und Kriechbruchs
— Verformung und Bruch teilchenverstärkter Legierungen
— Kriechen und Kriechbruch von Stählen
— Schadensakkumulation und Kriechrißwachstum
— Verformung und Bruch unter kombinierten Kriech-Ermüdungs-Bedingungen
— Hochtemperaturverhalten von Schweißungen
— Bestimmung von Kriechdaten und Restlebensdauerprognose
— Hochtemperaturverhalten von Konstruktionskeramik

Die 75 Einzelbeiträge überstreichen ein breites Spektrum, das von den mikrostrukturellen Schädigungsmechanismen bis zur ingenieurtechnischen Bewertung zeitstandgeschädigter Bauteile reicht. Große Beachtung finden neuartige bzw. eigenschaftsoptimierte metallische und keramische Werkstoffe, wobei die sog. Deformations-Schaubilder eine übersichtliche Zuordnung der verschiedenen Kriechmechanismen zur Temperatur und mechanischen Beanspruchung ermöglichen. Von besonderer Aktualität sind die Untersuchungen zum Kriechrißwachstum und zur Anwendung der Kriechbruchmechanik (C*-Parameter). Als charakteristisches Merkmal dieser Arbeiten kann die immer enger werdende Verbindung von experimentellen, analytischen und numerischen Methoden angeführt werden.

Obwohl der Monografiecharakter des — auch in seiner äußeren Form ansprechenden — Tagungsbandes durch einige Übersichtsbeiträge noch etwas stärker zur Wirkung hätte gebracht werden können, ist er für Festkörperphysiker, Festkörpermechaniker und Werkstoffwissenschaftler, die auf diesem Gebiet Grundlagen- oder Anwendungsforschung betreiben, von großem Gewinn.

H. Blumenauer

Crystal Research and Technology

Zeitschrift für experimentelle und technische Kristallographie

Edited by H. NEELS, Leipzig

Papers in Number 10, 1987

Original Papers

AKADEMIE-VERLAG · BERLIN

Continued from cover page 4

Short Notes

Book Reviews

Crystal Research and Technology
Volume 22 1987 Number 9

Contents

Original Papers

Continued on cover page 3

Crystal Research and Technology is indexed in Current Contents/PC & ES.

FSC
www.fsc.org
MIX
Papier | Fördert
gute Waldnutzung
FSC® C083411

Kolibri 360 GmbH
Ferdinand-Jühlke-Straße 7
99095 Erfurt, Deutschland
produktsicherheit@kolibri360.de